STATISTICS ON SPHERES

The University of Arkansas Lecture Notes in the Mathematical Sciences

Volume 6: Geoffrey S. Watson, *Statistics on Spheres*

The first five volumes in the Series were published by the University of Arkansas Department of Mathematics

Volume 1. G. G. Lorentz with a contribution by P. Nevai, *Problems in Approximation Theory*

Volume 2. James Glimm and Arthur Jaffe, *Probability Applied to Physics*

Volume 3. Kyoshi Iseki, *Non-Greek Mathematics*

Volume 4. W. A. J. Luxemburg, *Some Aspects of the Theory of Riesz Spaces*

Volume 5. Nathan Jacobson, *Structure Theory of Jordan Algebras*

VOLUME 6

STATISTICS ON SPHERES

GEOFFREY S. WATSON
Princeton University

A WILEY-INTERSCIENCE PUBLICATION
JOHN WILEY & SONS
New York Chichester Brisbane Toronto Singapore

Library of Congress Cataloging in Publication Data:

Watson, Geoffrey S.
Statistics on Spheres.

"A Wiley-Interscience publication."
Bibliography: p.
Includes index.
1. Mathematical statistics. 2. Sphere. I. Title.
QA276.W37 1983 519.5 83-1328
ISBN 0-471-88866-4

A Note from the Publisher

Wiley-Interscience is pleased to publish STATISTICS ON SPHERES, the sixth volume in the University of Arkansas Lecture Notes in the Mathematical Sciences. The first five volumes were published by the University. We look forward to publishing the Series and are proud to be associated with this collection of outstanding monographs.

Preface

The work of this author which is described in these notes was partially supported since 1959 by the Office of Naval Research on various contracts, most recently, N00014-79-C-0322.

It was motivated by questions from workers in the Earth Sciences and in Animal Behaviour although now it has an independent theoretical interest for me. In particular, E. Irving's enthusiasm for rock magnetism, and continental drift before it was accepted, gave me the reason for developing easy methods for Fisher's distribution and K. Schmidt-Koenig's interest in homing pigeons lead to my interest in goodness-of-fit tests. I have been fortunate in having M.A. Stephens and R.J.W. Beran as students and then as coworkers in these two topics on which they have done so much.

My interest was rekindled by invitations to speak on this topic from K.V. Mardia, whose work and book brought the subject to the attention of a wider circle of statisticians, and by encouragement from D.G. Kendall to describe the area as now I saw it. The kind invitation by the University of Arkansas to deliver and write up these lectures came at the right time and I am most grateful for it. Some of the writing and thinking was done while I was a visitor at the Universities of Aarhus, Cape Town and Melbourne.

I wish to state categorically that what follows is not a balanced text on "Statistics on Spheres" but an arrangement of material, mainly my current work, into five lectures. The sequence is not unnatural but it is incomplete in many ways, only some of which I know, at the moment, how to complete.

I would be glad to hear of errors and misprints which surely abound! But for help from Professors Louis Paul Rivest and Jon Wellner, and particularly from Professor Christopher Bingham, there would have been many more so I wish to express my gratitude to them.

These notes have been retyped in several places and I regret the uneveness of the type. I wish to thank Marilyn Cherkowsky and Monica Selinger.

G.S. Watson

Fine Hall, Princeton
August 1982

Contents

STATISTICS ON SPHERES

Chapter 1 Introduction

1.1 Scope

Statisticians mostly have to deal with data in the form of counts or of measurements or combinations of both. When many members of the same population are examined, the frequency distributions of counts, the histograms of measurements, usually take one of a small number of shapes which characterize the distribution of this property in the population. Thus, e.g. counts are often seen to have the binomial, Poisson distributions, etc., and measurements, on some scale, the Normal or Gaussian distribution, the chi-square distribution, etc..

If this is so, different populations may be contrasted by comparing the parameters (more generally functionals) of the distributions which describe them. The theory of statistical (point, interval) estimation provides ways of using the data from each population to obtain a good estimate of its parameters together with an estimate of its precision e.g. its variance. To get a feeling for whether estimates of parameters differ only because of the fluctuations of random sampling, not because the population parameters are different, the theory of testing statistical hypothesis may be used to provide tests.

The theory and practice of statistics then consists of exploratory work -- trying to find statistical regularities and relationships including the joint distribution of the measurements -- and, given such regularities, estimation of parameters and testing hypotheses about relationships between them. In the second endeavor, probability theory is essential. In

Science, statistical and probabilistic are almost synonymous adjectives. The distribution of data often supports theory about the processes that lead to it. Thus probabilistic (or stochastic, another synonym) models are of interest to the statistician; the simplest example is the argument which leads to the binomial distribution. Books on this topic usually have "Stochastic Processes" or "Applied Probability" in their titles.

In these notes we will be interested in measurements of directions (sometimes unsigned) i.e. in unit vectors or points on a sphere Ω_q in a space of arbitrary dimension q , $\mathbb{R}^q$. Most orientation data of course are in two or three dimensions. Examples will be given in the next two sections. From the above it is clear that (i) we will have to provide methods showing visually the distribution of the directions in a sample; (ii) we will need parametric probability distributions that fit the simpler data distributions encountered; (iii) we will be interested in stochastic processes that lead to these distributions; (iv) we will have to provide methods of estimation and testing for these distributions.

The only novelty is that our variables x will be constrained to the sphere Ω_q , and not allowed to vary over all of $\mathbb{R}^q$. The branch of statistics dealing with data (and random variables) in $\mathbb{R}^q$ is called "Multivariate Analysis". Much of this theory deals with Gaussian distributions in $\mathbb{R}^q$ for which we too will find great use in deriving estimation and testing methods. But they are not available for our basic data distributions since these must be wholly concentrated on the sphere Ω_q . However our subject matter is properly thought of as a special form of multivariate analysis solely designed for use with data defined on the sphere. From a mathematical point of view, it is tempting to consider other smooth manifolds embedded in $\mathbb{R}^q$ and not just the particular case of the sphere but we will resist this here.

It will be assumed that the reader is familiar with the elementary aspects of probability theory and statistics presented in most first courses in "Mathematical Statistics". An effort has been made to derive all results from first principles. The level of sophistication will however go up and down since some readers may want only an intuitive understanding in order to apply the methods while others will be primarily interested in the theoretical questions behind the methods.

In fact, for the most part, we stop just short of practical methods. In five lectures-essentially the five chapters-there simply was not enough time to separate the methods from the theory and to provide worked examples. The implied methods given below are all designed for "large" samples. We have no clear answers to the obvious question-for what finite samples, will these approximations be satisfactory. There are finite sample methods for some of the problems discussed. And there are many practical problems for which we have the methods but which could not be included for lack of time and space.

Chapter 1 gives practical examples that raise statistical problems and some data-analytical tricks. Chapter 2 studies the basic distribution, the uniform on Ω_q . Chapter 3 deals with both theoretical statistical reasons and applied probability reasons for the prevalence of certain non-uniform distributions on Ω_q . In Chapters 4 and 5, a large sample theory is given for the several-sample problem for two classes of distributions.

1.2 Data and distributions on a circle.

This writer was introduced to circular data by Klaus Schmidt-Koenig who was interested in the homing ability of pigeons. Homing pigeons, when released some distance from their home loft, climb in a spiral and then fly off. Their bearings, from the release point, as they vanish are recorded to the nearest 5^0. If a bird navigates by using its (internal) clock and the Sun, as mariners do, one would expect that a 2-hour day shift in its clock would set them off their true direction by $360^0/24=15^0$. Figure 1.1 shows the results of an experiment by Schmidt-Koenig (1972) using 93 birds whose clocks have not been shifted (the controls) and 99 whose clocks have been shifted by two hours (the experimentals). He has made a histogram of his data (using 5^0 intervals) but put it around a circle with the true home direction at the top of the circle. For fairly small data sets, sometimes distinct filled circles or triangles or dots are used for each data point instead of a box or bar of length equal to the number of circles or triangles. Sometimes the symbols are placed inside the circle. The recent book by Batschelet (1981) is brilliantly illustrated and shows many interesting data sets. The data in Fig. 1.1 is quite widely disperesed about some central direction or bearing and this dispersion looks about the same in each case. Furthermore the preferred or modal homing direction of the experimentals does seem to have shifted clockwise by about 15^0. Can one show that this is not due to wishful thinking or random fluctuations in bird behavior? How should one define the preferred or average or modal direction and the dispersion of the sample about it?

It is clearly dangerous, if the data are angles $\theta_1,\ldots,\theta_n$ to use
$\bar\theta = n^{-1}\sum_1^n \theta_i$, $s^2 = (n-1)^{-1}\sum_1^n(\theta_i-\bar\theta)^2$.

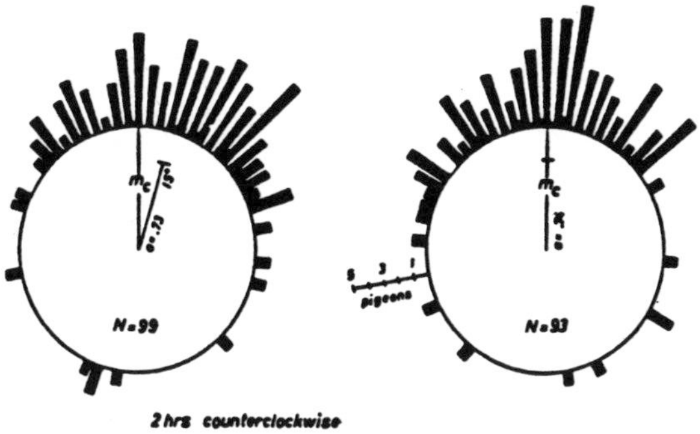

Fig.1.1 Initial orientation (circular diagrams) and homing performance (rectangular diagrams) of control birds and of experimental birds upon 2-hr shifts counterclockwise from 460 and 320 km S and from 280 km E to Wilhelmshaven, Osnabrück, Germany, and Durham, N. C.

(From K. Schmidt Koenig's article (pp 275-282) in "Animal Orientation and Navigation," published by NASA (SP262) and edited by S.R. Galler, K. Schmidt-Koenig, G.J. Jacobs and R.E. Belleville).

Suppose the data are $(1^o, 359^o)$. Then $\bar{\theta} = 180^o$, $s^2 = 2 \times 179^2$. If the data is plotted as in Figure 1.1, there is only one box of height two centered on $\theta = 0$! Had the data been written $(1^o, -1^o)$, we would get the more sensible answers $\bar{\theta} = 0$, $s^2 = 2 \times 1^2$. Had the origin been changed by 1° , the data would read $(2^o, 0^o)$ or $(2^o, 360^o)$ which again leads to distinct answers. These troubles disappear if one thinks of the data as unit vectors with components $(\cos\theta_i , \sin\theta_i)$, $i=1,\dots,n$. The vector resultant of the data is a vector with components $(\sum_1^n \cos\theta_i \sum_1^n \sin\theta_i) = r(\cos\hat{\theta} , \sin\hat{\theta})$ where $r^2 = (\Sigma\cos\theta_i)^2 + (\Sigma\sin\theta_i)^2$, $\tan\hat{\theta} = \Sigma\sin\theta_i / \Sigma\cos\theta_i$. By drawing the vector resultant of several data sets, one sees that $\hat{\theta}$ gives a sensible measure of the center of data of the type seen in Figure 1.1. Further if the data points are very dispersed, the length of the vector resultant will be small relative to n while if the data are concentrated it will be large -- in fact almost as big as n , its mathematical maximum attained only when all the data points are identical. Thus $n-r$ is a sensible measure of the dispersion of the whole sample about its estimated center and a good analogue of $\Sigma(x_i - \bar{x})^2$ for data points x_i on a <u>straight</u> line.

Thus $(n-r)/(n-1)$ is a possible analogue of $\Sigma(x-\bar{x})^2/(n-1)$ as a measure of the dispersion of data on a circle. This is further supported by the identity,

$$n-r = \Sigma \| x_i - \frac{\bar{x}}{\|\bar{x}\|} \|^2$$

where the norm of a vector z, $\|z\|$, is defined by $\|z\| = (z'z)^{\frac{1}{2}}$ and $x = n^{-1}\Sigma x_i$.

Alternatively if one imagines each data point as a unit mass, their center of mass would be at the point $n^{-1}(\Sigma\cos\theta_i, \Sigma\sin\theta_i) = a(\cos\hat{\theta}, \sin\hat{\theta})$. If the points are widely dispersed, this point will be near the center of the circle i.e. a will be nearly zero. If they are concentrated, it will be near the circle i.e. a will be near unity. Hence a is a measure of concentration so $1-a$ is a measure of dispersion. But $a=r/n$ so we arrive at the measure $(n-r)/n$ of dispersion -- essentially as before.

If we could make enormous experiments under constant conditions and measure the bearings more accurately, we could make smoother histograms on the circle. As n increases, it is better to make the boxes with decreasing bases and rescale the heights so the total area of all the boxes is unity. Then the tops of the boxes should tend to a function $f(\theta)$ such that, if θ is measured in radians and not degrees,

$$f(\theta) \geqslant 0 \ , \int_{\theta'}^{\theta''} f(\theta)d\theta = \text{proportion of the population of birds vanishing in the interval } (\theta', \theta'') \ , \tag{1.2.1}$$

$$\int_0^{2\pi} f(\theta)d\theta = 1 \ . \tag{1.2.2}$$

Thus $f(\theta)$ is a probability density function. Any function satisfying (1.2.1) and (1.2.2), could conceivably appear. Figure 1.1 suggests that the population density might be symmetric about some bearing θ_0 say and so f might be a function of $\cos(\theta-\theta_0)$. A very convenient (and plausible) suggestion for Figure 1.1 is

$$f(\theta) = \frac{\exp \kappa \cos(\theta-\theta_0)}{2\pi\, I_0(\kappa)} \tag{1.2.3}$$

where $I_0(\kappa)$ is a modified Bessel function, one of whose definitions is

$$I_0(\kappa) = (2\pi)^{-1} \int_0^{2\pi} \exp \kappa \cos\phi \, d\phi = \sum_{j=0}^{\infty} \frac{1}{(j!)^2} \left(\frac{\kappa^2}{4}\right)^j , \tag{1.2.4}$$

so that (1.2.1) and (1.2.2) are satisfied. If κ is zero, $f(\theta) = 1/2\pi$, the distribution is uniform and has no preferred direction. As κ increases from zero, $f(\theta)$ peaks higher and higher about θ_0 , so κ is a concentration parameter.

Similar experiments with turtles and other beasts (see e.g. Batschelet (1981)) tend to show a smaller peak 180^0 away from the main peak so one is tempted to suggest the variant of (1.2.3) ,

$$f(\theta) = C(\kappa_1,\kappa_2) \exp (\kappa_1 \cos(\theta-\theta_0) + \kappa_2 \cos 2(\theta-\theta_0)) \tag{1.2.5}$$

where $C(\kappa_1,\kappa_2)$ is chosen so that (1.2.2) is true, and κ_2 is smaller than κ_1 . This density is bimodal. More generally, <u>any</u> practical density symmetric about θ_0 can be represented by a density proportional to

$$\exp \sum_1^M \kappa_m \cos m(\theta-\theta_0) \tag{1.2.6}$$

by including enough terms in the sum. The use of the exponential ensures that the densities be non-negative and has (as will be seen later) other mathematical benefits not shared by densities of the form

$$\frac{1}{2\pi}\left(1 + \sum_{1}^{\infty} \lambda_m \cos m(\theta-\theta_o)\right), \tag{1.2.7}$$

which will be non-negative if the λ_m are small enough. A general form for a not necessarily symmetric distribution about θ_o would have a density proportional to

$$\exp \Sigma\, \alpha_m \cos m(\theta-\theta_o) + \beta_m \sin m(\theta-\theta_o)\ . \tag{1.2.8}$$

With data $\theta_1,\ldots,\theta_n$ on a circle, it is natural to use trigonometric rather than power moments. Generalizing (1.2.7), suppose that

$$f(\theta) = \frac{1}{2\pi}\sum_{-\infty}^{\infty} c_m \exp im\theta \qquad (i = \sqrt{-1})\ .$$

Since $\displaystyle\int_0^{2\pi} \exp im\theta\, d\theta = \begin{cases} 0 & , m\neq o \\ 2\pi & , m=o \end{cases},$

$$E(e^{-im\theta}) = \int_0^{2\pi} e^{-im\theta} f(\theta)d\theta$$

$$= c_m \quad ,$$

so that

$$E\left(\frac{1}{n}\sum_{j=1}^{n} \exp -im\theta_j\right) = c_m \quad .$$

Since the integral of $f(\theta)$ must be unity, $c_0 = 2\pi$. Thus we can estimate the coefficients c_m in the series for $f(\theta)$ by

$$\hat{c}_m = \frac{1}{n}\sum_{1}^{n} \exp - im\,\theta_j$$

Of course f is real so that $\bar{c}_m = c_{-m}$, $\bar{\hat{c}}_m = \hat{c}_{-m}$. But the obvious estimator $\hat{f}$ of f defined by

$$\hat{f}(\theta) = \frac{1}{2\pi}\sum_{-\infty}^{\infty} \hat{c}_m \exp im\theta$$

is unsatisfactory and one needs to put in convergence factors a_m, $a_m \to o$ as $m \to \infty$ to decrease the effect of the higher terms. This will lead to the estimator

$$f^*(\theta) = \frac{1}{2\pi}\sum_{-\infty}^{\infty} \hat{c}_m a_m \exp im\theta = \frac{1}{n}\sum_{1}^{n} a(\theta - \theta_j),$$

where $a(\theta) = \frac{1}{2\pi}\sum_{\infty}^{\infty} a_m \exp im\theta$. The estimator $f^*(\theta)$, unlike $\hat{f}$, is biased but is smoother than $\hat{f}$ and will have a smaller variance. A suggestion for how to choose the a_m's is given in Watson (1969). For an understanding of the numerical aspects of Fourier series, see Hamming (1973). These ideas lead to one way of making a non-parametric estimate of a density function. Another method will be suggested in the next section. This topic is of general interest - see e.g. Tapia and Thompson (1978). All methods stem from the desire to draw a smooth curve through a histogram and to avoid the arbitrary choice of class intervals needed to make a histogram.

In practice we will want to use a density which, while adequately fitting the data, has few parameters. There is no clear answer to this problem which can be restated in several ways. A related problem, which will be discussed several times below, is to devise tests for whether a sample could reasonably have come from a population with a specified density, e.g. (1.2.3) -- the goodness-of-fit problem. A particular case of this is often used with directional data.

A recent experiment, Gould and Able (1981), tried to detect the ability of humans to sense the direction of the earth's magnetic field. While the design was actually more complicated, suppose subjects had been spun, blindfolded, in a chair, and then asked to point to north. One would guess that the indicated bearings would be uniformly distributed around the circle. This is the <u>null hypothesis</u> to be tested. Section 2.5 will be devoted to this problem and its generalizations but we can here hazard a guess that a sensible test would reject the null hypothesis only if r, the length of the sample resultant, is "too" large. To find out where to draw the line, we need to know the probability distribution of r when the data has, in fact, been drawn from a uniform distribution -- a problem solved in Chapter 2.

Suppose that the parametric form of the population has been agreed to, e.g. for the data in Figure 1.1 that we are to use (1.2.3). Then a comparison of the control and experimental birds means a comparison of the parameters κ and θ_0 for each population. For this we may use the standard methods of mathematical statistics. Details are given in Chapter 4.

If the parametric form is not agreed to and the null hypothesis is that resetting the birds clocks by 2 hours has left their homing ability unaltered, it may be tested by the methods of Chapter 4 if the sample sizes are large.

Data on a circle often has a quite different origin. Von Mises (1918) first suggested the distribution (1.2.3) in a problem to do with atomic weights which must be integers but the experimental determinations are not. This statistical problem would now be treated differently (see e.g. Hammersley (1950)) but von Mises treated the fractional parts of the measurements as distributed on a circle of unit perimeter. Much data has been collected on the time of day of human births, suicides, etc.. This would naturally be shown, as in Figure 1.1 where the circle is now a clock face divided into 24 hours. Similarly the dates of deaths from a specific cause might be put on a circle marked from 1 to 365. In each case, the histogram describes a population from which the data is supposedly a random sample. The same plots may be made of wind directions but here one must beware of the correlation between successive observations, seasonal changes, etc..

Geology provides many further examples but these will be emphasized in the next section.

Alternative methods of displaying directional data in the plane are shown in Figure 1.2. It will be noted that Figure 1.2(c) has a special form because axes or unsigned directions are involved. The axis of a fold is a line. If these lines are drawn through the center of the circle, they cut the circle twice. Both points have been used in forming this version of the histogram where instead of boxes being placed on the

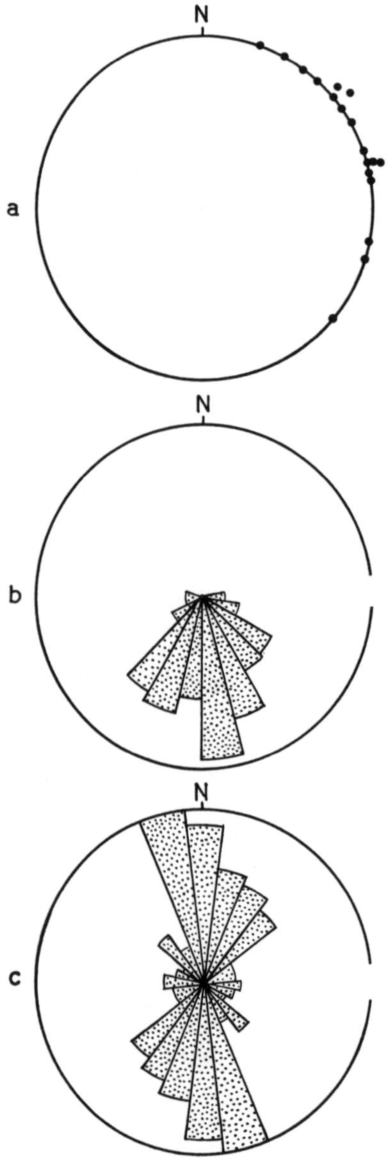

Fig. 1.2 (a) Raw data plotted on circle; (b) rose diagram of some palaeocurrent directions; (c) rose diagram of some axes of slump folds. (Redrawen from Potter & Pettijohn (1963).)

perimeter, sectors of radii proportional to the frequency have been drawn out from the center. The radii in Figure 1.2(b) are also proportional to the frequency. The area of a sector is $\frac{1}{2}(\text{radius})^2(\text{angle})$ so with this choice, areas are not proportional to the fractions of observations as with the areas in Figure 1.1. If the radii are made proportional $(\text{frequency})^{\frac{1}{2}}$, this problem is overcome and another benefit is derived. If the frequencies are small, the variance of $(\text{frequency})^{\frac{1}{2}}$ is more nearly constant so it is less easy to misinterpret "bumps" in the rose diagram.

Sometimes along with each direction there is paired a scalar, for example wind bearing and speed. In this case it seems natural to use the actual vector velocity, but Jensen (1981) has argued for this separation. Such a population no longer fits on a circle or sphere and so is slightly outside our scope.

1.3 Data and distributions on a sphere.

This topic was begun in Arnold's (1941) unpublished Ph.D. dissertation. He was motivated by James Bell's dissatisfaction with Sander's (1930) statistical treatment of the preferred direction for the optical axes of crystals in rocks and Krumbein's (1939) study of the orientation of the long axes of pebbles in glacial tills. Arnold, both for the circle and the sphere, considered two mathematical methods for deriving probability distributions for directions and axes, a topic which will be treated in detail in Chapter 3. One of the distributions he investigated was the 3-dimensional extension of the von Mises density (1.2.3), now known as the Fisher distribution.

Denote by x and μ, unit vectors in $\mathbb{R}^3$, and let κ be a non-negative real number. Then

$$f(x) = \frac{\kappa}{4\pi \sinh\kappa} \exp \kappa \mu' x , \qquad (1.3.1)$$

where μ' is the transpose of μ, is a probability density on the sphere. If orthogonal vectors e_1, e_2, e_3 with $e_3 = \mu$, and spherical polar coordinates θ, ϕ are introduced so that $x = \sin\theta\cos\phi \, e_1 + \sin\theta\sin\phi \, e_2 + \cos\theta \, e_3$, the density (1.3.1) has the form

$$f(\theta,\phi) = \frac{\kappa}{4\pi \sinh\kappa} \exp \kappa \cos\theta . \qquad (1.3.2)$$

The area element on the unit sphere is $\sin\theta d\theta d\phi$. The density (1.3.1.) is rotationally symmetric about μ, its mode. It is convenient to call μ the <u>polar</u> or modal vector. From (1.3.2) we see that ϕ is uniformly distributed on $(0,2\pi)$ and that the density of θ is given by

$$\frac{\kappa}{2\sinh\kappa} \exp(\kappa\cos\theta) \sin\theta .$$

The distribution (1.3.1) or (1.3.2) was introduced by Fisher (1953) for the analysis of paleomagnetic data. In paleomagnetic studies, the direction only of the remanent magnetism of ancient rocks is used because the strength may have decayed for various reasons. Fisher, like Arnold, discussed the maximum likelihood estimation of κ and μ but went on to consider fiducial distributions for κ and μ and provided some basic sampling distributions. The modern interest in the subject matter of this book (and this writer's) dates from Fisher's basic paper. The generalization and initial discussion of statistical inference of (1.3.1) to unit vectors in $\mathbb{R}^q$ was made in Watson and Williams (1956).

Suppose that n unit vectors $x_1,\ldots,x_n$ have been independently drawn from (1.3.1). The logarithm of their likelihood is, to a constant,

$$n(\log\kappa - \log\sinh\kappa) + \kappa\, \mu' X , \tag{1.3.3}$$

where $X = x_1+\ldots+x_n$, the sample resultant. The values $\hat{\kappa}$ and $\hat{\mu}$ which maximize (1.3.3) are seen to be defined by

$$\hat{\mu} = X/ \|X\| , \tag{1.3.4}$$

$$\coth\hat{\kappa}-\hat{\kappa}^{-1} = \|X\| /n , \tag{1.3.5}$$

where $\|X\|$ denotes the length of X which must lie in $(0,n)$. The left hand side of (1.3.5) increases monotonically from zero to unity as $\hat{\kappa}$ increases from zero to infinity so (1.3.5) always has a unique solution. κ is a concentration parameter. Now $\coth\kappa - \kappa^{-1} = 1-\kappa^{-1} + O(\kappa^{-2})$ so that for $\|X\|$ near n one has approximately

$$\hat{\kappa} = \frac{n}{n - \|X\|} = \frac{1}{1-\bar{x}} \quad , \quad \bar{x} = \frac{X}{n} \quad . \tag{1.3.6}$$

Since $1/\hat{\kappa}$ has the sense of dispersion and is given by $(n - \|X\|)/n$, we see that the maximum likelihood estimates of κ and μ for (1.3.1) support the intuitive discussion given in Section (1.2) which is equally applicable here.

In fact the density (1.3.1) had arisen in Langevin's (1905) statistical mechanical discussion of magnetism. If a dipole of moment m, in a parallel magnetic field of strength H in the direction μ , is excited by molecular motion, the Boltzman distribution for its orientation is (1.3.1) with κ=mH/kT, where T is the absolute temperature and k is Boltzman's constant. Of course Langevin was not involved in data analysis in our sense but the function $\coth\kappa-\kappa^{-1}$ which arises in the maximum likelihood estimation of κ is known in Physics as the Langevin function, and its properties continue to be of interest there. (See e.g., Dyson, Lieb, and Simon (1978).

It would therefore seem to settle priority disputes if (1.3.1) in any number of dimensions were known as the Langevin distribution.

As an axial distribution on the sphere, Arnold gave, as the analogue of (1.3.1), a density proportional to $\exp\kappa|\mu'x|$, a suggestion later made by Selby (1964). This density is axial because the density is the same at the point x as it is at the point $-x$. A more manageable density of the same type is proportional to

$$\exp \kappa \, (\mu' x)^2 \ , \qquad (1.3.7)$$

as suggested and used by Scheidegger (1965) and Watson (1965). If $\kappa > 0$, the density is symmetric bimodal and if $\kappa < 0$, the density peaks around the great circle made of points x, 90^0 away from the mode $\pm\mu$. It is rotationally symmetric about μ. Bingham (1974) suggested a generalization with density proportional to

$$\exp\left(\kappa_1(\mu_1' x)^2 + \kappa_2(\mu_2' x)^2 + \kappa_3(\mu_3' x)^2\right) = \exp x' K x \qquad (1.3.8)$$

where

$$K = \kappa_1\mu_1\mu_1' + \kappa_2\mu_2\mu_2' + \kappa_3\mu_3\mu_3' \ , \qquad (1.3.9)$$

the spectral form for K since in (1.3.8), μ_1, μ_2, μ_3 are three orthogonal unit vectors.

Rather than develop more formulae for densities on the sphere as in Section 1.2 by using series of (surface) spherical harmonics instead of Fourier series, a topic which arises naturally in Chapter 3, we turn to more examples of, and the specification and display of, directional data in three dimensions. For the Earth Sciences, Watson (1970) has given an explanation of the technical terms and brief accounts of oriented quantities occurring particularly in sedimentary and structural geology, petrofabrics and palaeomagnetism.

In sedimentary geology, there are various indications in rocks of ancient direction of flow of water (e.g., flute marks), ice (e.g., striations) and, much more rarely, of wind -- a classic reference is Potter and Pettijohn (1963). A bedding plane is sometimes given a directed normal by pointing it in the direction of more recent deposition -- the younging or facing direction. For an indication of orientation problems in Structural Geology, the book by Ramsey (1967) may be consulted. Even within the Earth Sciences directions are sometimes specified differently. A directed line, such as a glacial striation, which is measured in the field may be specified in a number of ways and the measurements are usually made with a Brunton Compass, which in addition to a magnetic needle has a level bubble and, essentially, a protractor. The angle between two vertical planes, through the sampling point, one containing the vector and one containing the magnetic north pole is the *trend* angle. It is given in degrees east of or clockwise from the north. It is sometimes called the *azimuth* or *declination*. The "*dip*" angle is the angle between the vector and the horizontal plane, taken as positive if the vector is below the horizontal plane. The "*dip*" is sometimes called the *inclination* or *plunge*. Thus a vector with trend t and dip d would have direction cosines on axes vertically downwards, northwards, eastwards of $\sin d$, $\cos d \cos t$, $\cos d \sin t$, respectively. Further specifications are given below.

The orientation of a plane is invariably defined by its *strike* and *dip*. The bubble allows the determination of a horizontal line in the plane. Choose the smaller of the two possible angles this line makes with north as the *strike* of the plane. Thus strikes run from -90^0 (W) to $+90^0$ (E). The *dip* of the plane is the dihedral angle with the horizontal plane; to avoid ambiguity dips must be described as "to the East" or "to the West". If the plane has a facing, a directed normal or pole may be defined. If a plane has strike s, and the up-face dips down to the west by an amount $\tilde{d}$, the direction cosines of the normal of the up-face relative to downwards, northwards, eastwards axes are $(-\cos \tilde{d}, -\cos s \sin \tilde{d}, \sin s \sin \tilde{d})$.

Directed lines are often in planes (e.g. striations). If the horizontal line in the plane and its strike is found, the angle between the directed line and the strike line (given its northern sense) is the *pitch* of the directed line. Pitch may be determined more accurately than trend because it is not necessary to estimate the imaginary vertical plane through the lineation. *Strike, dip and pitch* specify the direction uniquely. *Trend and plunge* or *declination and inclination* or *azimuth and "dip"* (pairing the words as they are used in different subjects) do so too. Note that dip is used in two senses so that we have used "dip" and dip; we will henceforth drop the azimuth and "dip" terminology. Declination and inclination is a usage restricted to magnetism.

There are six combinations {(Strike, dip, trend), (Strike, dip, pitch), (Strike, dip, plunge), (Strike, trend plunge), (Dip, trend, plunge), (Dip, trend, pitch)} which could be used to specify the orientation of a directed line in a plane. The combination which gives the greatest accuracy depends on the practical situation. The measurement errors in some of these quantities are correlated and dependent on the values of other quantities, e.g. when a plane is almost horizontal (low dip) its strike is hard to determine. While these problems are well known and intuitive "solutions" are part of the practical training of geologists, there do not appear to be any formal studies in print.

The various measurements mentioned above also suffice to orient undirected lines and planes which have no preferred "facing direction".

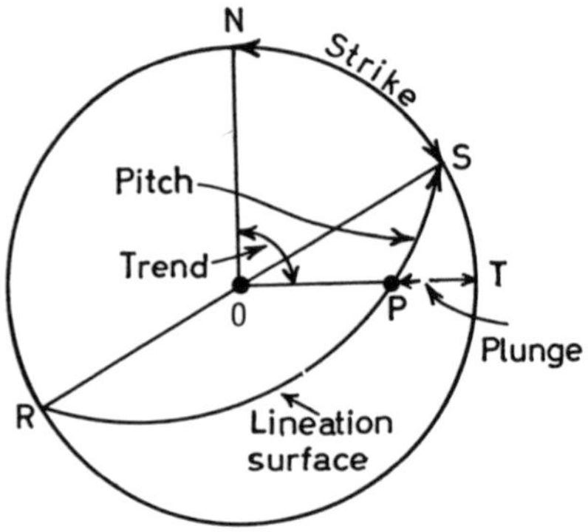

Fig. 1.3 The curve RPS is the intersection of a plane (the lineation surface) with the lower hemisphere, the plane of the paper being the horizontal, the view being from below the sphere. The orientation of a line in this plane is shown by OP. A vertical plane through the line cuts the hemisphere along the line OPT; this plane is, in practice, judged by eye. In terms of this Figure, the angles of pitch, plunge, strike, and trend are defined.

To look at directional data in three dimensions, some form of plane projection will usually be employed. The ordinary equal area projection is derived as follows. Suppose points on the unit sphere are specified by their spherical polar coordinates (θ,ϕ) $0 \leqslant \theta \leqslant \pi$, $-\pi < \phi \leqslant \pi$, the natural terminology for mathematics. The area element on the sphere is $\sin\theta d\theta d\phi$. Suppose the point (θ,ϕ) is mapped into a point $(\rho\cos\phi , \rho \sin\phi)$ on the plane through the equator $\theta = \pi/2$. Since the area element in the plane is $\rho d\rho d\phi$ (see Figure 1.4), to obtain an equal area projection of the sphere on a disc in the equatorial plane we must therefore have

$$\rho d\rho d\phi = \sin\theta d\theta d\phi \quad , \tag{1.3.10}$$

so that

$$\frac{d\rho^2/2}{d\theta} = \sin\theta \quad . \tag{1.3.11}$$

The point $\theta=0$ should correspond to $\rho=o$, the center of the disc. Integrating the differential equation (1.19), we find

$$\rho^2/2 = 1-\cos\theta = 2\sin^2\theta/2$$

so that a point (θ,ϕ) on the sphere corresponds to a point (ρ,ϕ) in the plane by

$$\rho = 2 \sin \theta/2 \ , \ \phi=\phi \ , \ \theta \geqslant 0 \quad . \tag{1.3.12}$$

The projection described by (1.3.12) is ordinarily used to show only the points on the upper hemisphere. They all lie on a disc of radius $2^{\frac{1}{2}}$. Points on the lower hemisphere are shown, with a different symbol, on the

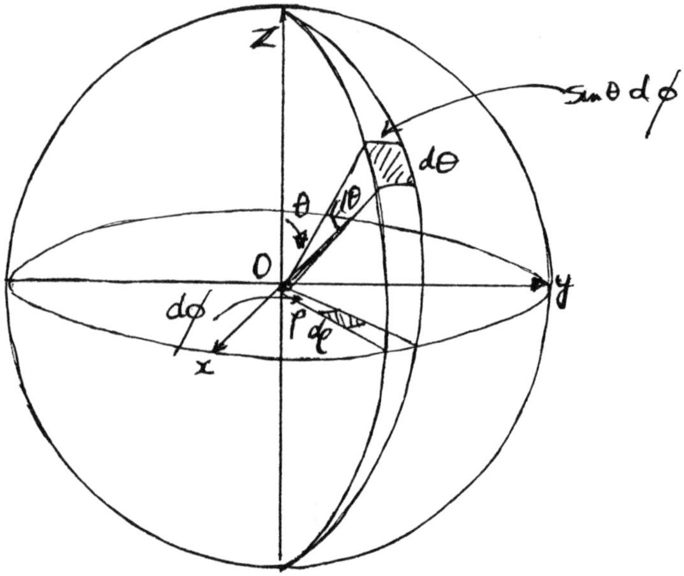

Fig. 1.4. Spherical polar coordinates (θ,ϕ). Equal area projection equates area element on upper hemisphere with area element in xOy plane.

disc with positions $\rho=2\sin(\pi-\theta)/2$, $\phi=\phi$. Graph paper allowing direct plotting is called a Schmidt net. While this method preserves the density of the points, the shapes of any clusters will be changed, the more so as θ moves away from zero. Thus if all the points lie in some hemisphere it is usual to rotate the data, so that it is all seen and the "center" of the points is near $\theta=0$. Methods for rotating data are discussed below - see e.g. (1.3.15).

An alternative method, pointed out to me by Christopher Bingham, is to use (1.3.12) for $0 \leqslant \theta \leqslant \pi$ which puts the lower hemisphere (excluding the south pole) into the annulus $\sqrt{2} \leqslant \rho \leqslant 2$. However the distortion of the lower hemisphere is considerable.

One could also make an equal area projection onto a rectangle by setting

$$dudv = \sin\theta d\theta d\phi$$

so that

$$\left.\begin{array}{ll} u = 1 - \cos\theta , & 0 \leqslant u \leqslant 2 \\ v = \phi , & 0 \leqslant v \leqslant 2\pi \end{array}\right\}$$

Either of these methods is ideal for computer plotting.

The stereographic projection, which preserves *angles*, is made by the geometric construction shown in Figure 1.5. The projection is unbounded, shows every point on the sphere, and the shapes of small clusters remain unchanged though their sizes are distorted, the more so as their position approaches the point of projection, N. This projection is always used by crystallographers. Hobbs, Means and Williams (1976) contrast the two projections.

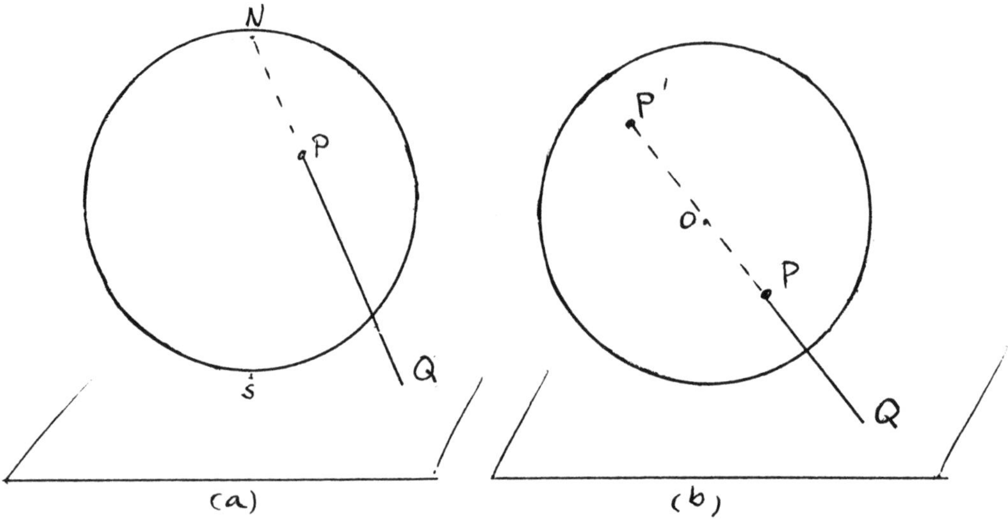

Fig. 1.5. In (a), imagine a sphere with the south pole setting on a plane. A ray from the north pole through a point P takes it to its stereographic projection Q .

In (b), the ray must go through the center O of the sphere so two antipodal points P´ are identified with the same point Q in the plane of projection.

If the point O in Figure 1.5 is used as the center of projection, antipodal points on the sphere will be identified with the same point on the plane of projection. This leads to a representation of the projective plane -- see e.g. Coxeter (1961) -- a fact mentioned by Bingham (1974).

For data analysis we will use only the equal area projection. The data in Figure 1.6 show some of the diversity obtained in the Earth Sciences.

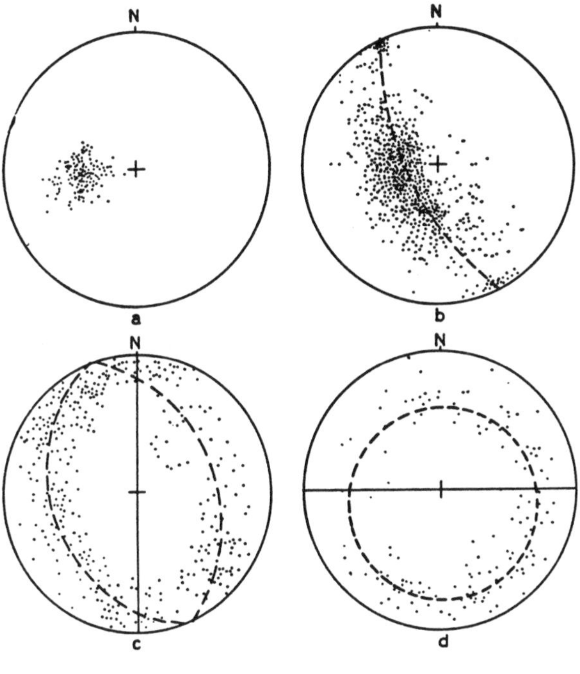

Fig. 1.6. Patterns of preferred orientation (median girdles show by broken lines). (*a*) Maximum (symmetric): 150 lineations from Loch Leven, Scottish Highlands. (*b*) Girdle: 1000 poles of foliation from Turoka, Kenya. (*c*) "Crossed girdle": 390 [0001] of quartz from quartzite, Barstow, California. (*d*) Small circle or "cleft" girdle: 140 [0001] of quartz from Orocopia schist, California.

(Redrawn from Fig. 3-7 of Turner & Weiss's "Structural Analysis of Metamorphic Tectonics," McGraw-Hill, 1963.)

Only the data in Figure 1.6(a) could conceivably be described by the density (1.3.1). The directions of magnetization of rock specimens sampled in the same rock formation often have an appearance similar to Figure 1.6(a). Such a cluster is well described by the modal vector $\hat{\mu}$ of (1.3.4) and the concentration parameter $\hat{\kappa}$ given by (1.3.5) (or even (1.3.6)). The other distributions shown in Figure 1.6 are not.

Since these measures were also derived (intuitively) by analogy with the center of mass, one wonders where an analogy with the moment of inertia matrix of the data will be helpful. Imagine then unit masses at each of the data points $x_1,\ldots,x_n$ on the sphere. The moment of inertia of this set of mass points about an axis a , $\|a\|=1$, is simply $\sum_1^n \|x_{i\perp a}\|^2$, where $x_{i\perp a}$ is the part of x perpendicular to a . By Pythagoras,

$$\sum_1^n \|x_{i\perp a}\|^2 = \sum_1^n \|x_i\|^2 - (a'x)^2$$

$$= n - a'(\sum_1^n x_i x_i')a . \qquad (1.3.13)$$

Those with physical science training will learn something about the distribution of the points by knowing the moment of inertia about various axes, a .

It is however simpler mathematically and computationally to examine, equivalently, the extreme values of $a'M_n a$ where

$$M_n = \frac{1}{n}\sum_1^n x_i x_i' , \qquad (1.3.14)$$

the second moment matrix of the data. This symmetric matrix is non-negative since $a'M_n a$ equals $n^{-1}\sum_1^n (a'x_i)^2$. Its real non-negative eigenvalues $\hat{\lambda}_1 \leqslant \hat{\lambda}_2 \leqslant \hat{\lambda}_3$ add to unity since their sum is the trace of M_n and trace $M_n = n^{-1}\,\Sigma$ trace $x_i x_i' = n^{-1}\,\Sigma$ trace $x_i' x_i = n^{-1} n$. Associated with the eigenvalue $\hat{\lambda}_i$ is the eigenvector $\hat{\mu}_i$ -- in fact, $M_n \hat{\mu}_i = \hat{\lambda}_i \hat{\mu}_i$.

Since $\|\hat{\mu}_i\|^2 = \hat{\mu}_i'\mu_i = 1$ by convention, $\hat{\lambda}_i = \hat{\mu}_i' M_n \hat{\mu}_i$. M_n , and the pairs $\hat{\lambda}_i, \hat{\mu}_i$ are easily computed from the data.

If the points are in two small antipodal clusters, $\hat{\lambda}_3$ will be large, the axis of clusters will be $\hat{\mu}_3$, and $\hat{\lambda}_1, \hat{\lambda}_2$ will be small. If the points are fairly uniformly arranged about a great circle, the eigen vector associated with $\hat{\lambda}_1$ will be normal to the plane of this great circle and $\hat{\lambda}_2$ and $\hat{\lambda}_3$ will be roughly equal and larger than $\hat{\lambda}_1$. If the points are fairly uniformly distributed over the sphere, $\hat{\lambda}_1, \hat{\lambda}_2, \hat{\lambda}_3$ will be roughly equal to each other and so to 1/3.

In practice we must look at the $\hat{\lambda}_i$ and infer facts about the distribution of the points. Observe that all the 2^n sets of vectors $\pm x_1, \pm x_2, \ldots, \pm x_n$ have the same $\hat{\lambda}_i$ and $\hat{\mu}_i$. Now that computers with visual displays have made the plotting and rotation of spherical data rapid and easy, this method of data analysis is less important. To rotate a data vector x , we need to be able to generate suitable orthogonal matrices H . The matrix

$$H_3(\alpha_3) = \begin{bmatrix} \cos\alpha_3 & -\sin\alpha_3 & 0 \\ \sin\alpha_3 & \cos\alpha_3 & 0 \\ 0 & 0 & 1 \end{bmatrix} \qquad (1.3.15)$$

makes a rotation of α_3 in the e_1-e_2 plane i.e. about the e_3 axis. $H_1(\alpha_1), H_2(\alpha_2)$ are defined similarly. The application of a suitably chosen sequence of these matrices will move a data cluster anywhere, and this is easier at a terminal than the classical rotation matrix using Euler angles -- for formulae see e.g. Messiah (1966). To rotate a point x to a point y, the orthogonal (and symmetric) matrix

$$H = \frac{(x+y)(x+y)'}{1+x'y} - I$$

may be used. It is easy to verify that $Hx=y$.

Marsden's (1979) catalogue of the orbits of all known comets provides an interesting data set. A directed normal to their orbital planes can be found using the right hand rule -- fingers pointing in the direction of the comet's motion, thumb indicating the normal. Using a coordinate system with the e_3 axis perpendicular to the plane of the ecliptic, the normals to orbital planes with only a few exceptions point upwards. Figure 1.7 shows the 659 data points. The cluster near the origin indicates that many comets move in the same sense in or near to the plane of the ecliptic. This fascinating data deserves more serious analysis than that given below which is based upon an undergraduate thesis by Elizabeth Ryder with additional checking by Javier Cabrera. If only the comets with periods greater than 1,000 years are plotted, we get Figure 1.8. The remaining comets plot as shown in Figure 1.9.

The 505 long period comets have normals that seem to be uniformly distributed over the hemisphere. The shorter period comet normals seem to have a distribution that might be fitted by (1.3.1), the Langevin distribution.

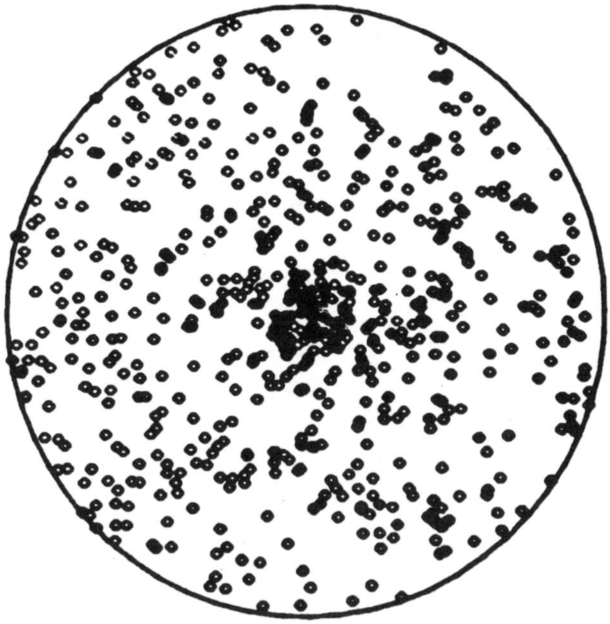

Fig. 1.7. An equal area plot showing 658 of the directed normals to the orbits of comets listed in Marsden's Catalogue (1979).

The eigenvalues of M_n for the data in Figure 1.8 are 0.284356, 0.330377, 0.385267, all similar, in agreement with the picture. If these comets were a sample from a population of comets with truly uniformly distributed normals, would we expect to see often such deviations of the λ_i from 1/3? To test the null hypothesis of uniformity it is natural to compute $\sum_1^3(\hat{\lambda}_i-1/3)^2$. In Chapter 2, we prove Bingham's (1974) result that, when the number n in the sample is large,

$$\frac{15n}{2}\sum_1^3(\hat{\lambda}_i-1/3)^2 \sim \chi_5^2 . \qquad (1.3.16)$$

In our case the left hand side of (1.3.16) is computed to be 19.3337. (Since trace $M_n^2 = \sum_1^3 \hat{\lambda}_i^2$, one does not have to compute the $\hat{\lambda}_i$ to evaluate the statistic.) Now a chi-squared random variable with 5 degrees of freedom only exceeds 15.09 with probability equal to 0.01 or 1% of the time. Hence this deviation is sufficiently rare in sampling from a uniform distribution of directions to make us doubt this assumption, despite the picture Figure 1.9!

Do the normals of the 153 shorter period comets look like a random sample from some Langevin distribution? The eigenvalues are .096930, 0.134654, and 0.768416. The last is larger than the others because of the cluster at or near the origin. But they don't help with our question. The length of the vector resultant $\|X\| = 99.4$, so $\hat{\kappa} = 2.77$ by solving (1.3.5). The approximation (1.3.6) gives $(n-1)/(n - \|X\|) = 2.8$. To answer the question raised above, we could divide the spherical surface into regions and use the fitted distribution to predict how many points we should expect in each region and complete the chi-square goodness of fit statistic, $\Sigma(O-E)^2/E$. This is neither very easy nor satisfactory.

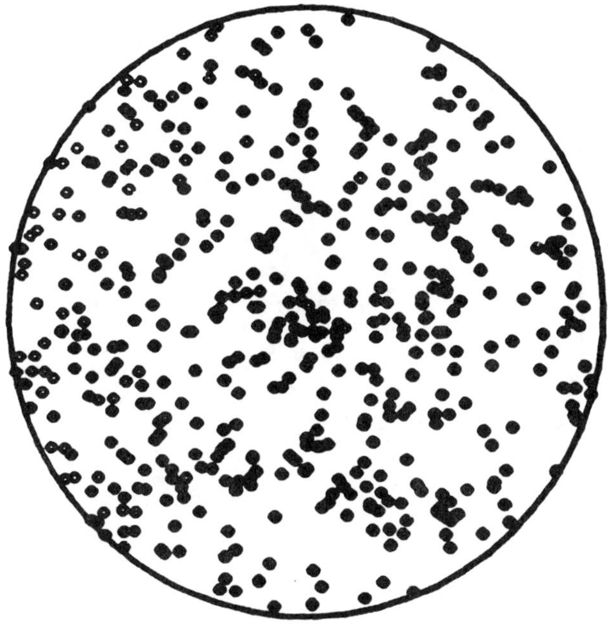

Fig. 1.8. An equal area plot of the normals to orbits of 505 comets with periods greater than 1000 years.

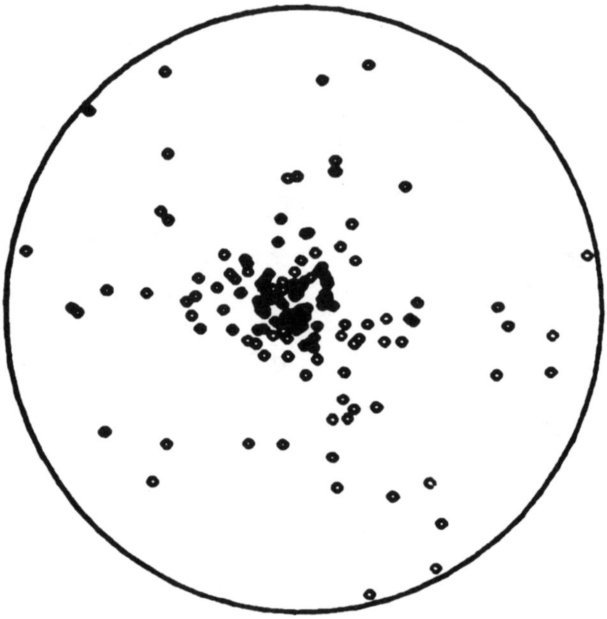

Fig. 1.9. An equal area plot of the normals to the orbits of 153 comets with periods of less than 1000 years.

If κ were a little larger than it appears to be, we could calculate the coordinates $(\hat{\theta}_i, \hat{\phi})$ of each data point where the $\hat{\theta}_i$ are measured away from $\hat{\mu}$ and then safely use the following strategy. If we really knew κ and μ the ϕ_i should be uniform on $(0,2\pi)$ and the θ_i have the density

$$\frac{\kappa}{2\sinh\kappa} \exp(\kappa\cos\theta) \sin\theta \ . \qquad (1.3.17)$$

Thus $v = \kappa(1-\cos\theta)$ has a density $\exp(-v)(1-\exp{-2\kappa})$ on $(0,2\kappa)$ and $u = \exp{-v}$ the density $(1-\exp{-2\kappa})$ on $(\exp{-2\kappa}),1)$. If κ is large, we see approximately, that $\kappa(1-\cos\theta_i)$ should have the standard exponential density on $(0,\infty)$, or equivalently, $\exp{-\kappa(1-\cos\theta)}$ should be uniformly distributed on $(0,1)$. Ignoring the effect of estimation we might then examine $\hat{\phi}_1,\ldots,\hat{\phi}_n$ for uniformity on $(0,2\pi)$ and $\exp{-\hat{\kappa}(1-\cos\hat{\theta}_1)},\ldots,\exp{-\hat{\kappa}(1-\cos\hat{\theta}_n)}$ for uniformity on $(0,1)$. Graphical methods include the stem-leaf (see Tukey (1977)) or P-P or Q-Q techniques.

In our case $\hat{\kappa}$ is too small for this procedure to be obviously safe. However the $\hat{\phi}_i$'s give no suggestion for non-uniformity but the $\exp{-\hat{\kappa}(1-\cos\hat{\theta}_i)}$ are <u>very</u> non-uniform. Figure 1.10 shows P-P and stem-leaf plots. The P-P plot should be approximately a straight line and the stem-leaf have rows of roughly equal length. Both tell the same story. There are far too many points near the center and very far from it for a sample from a Langevin distribution. The evidence is so strong that we feel it is not masked by the rough method.

With large samples of directions, one often feels the need for a contour map of the density of the data. See, for example, Figure 1.11. Thus one wants to see a projection of a contoured smoothed probability density estimator. This topic arose in the last section and here we show another method that might

```
00|2367555
02|2345781
04|12268013477889
06|27788856778999
08|455556812356
10|1288899
12|012282444569
14|127903458
16|357126
18|22881244559
20|246773367
22|34
24|03770145
26|223349023
28|018136
30|01459
32|078947
34|5667775689
```

Fig. 1.10a

```
00:00000001111111112234
00:77788
01:0334
01:58
02:014
02:679
03:11
03:
04:04
04:59
05:12
05:6
06:245
06:6777899
07:013
07:55888
08:113444
08:5566667788899
09:112233334444444
09:5555556666666677777777778888888888899999999999999999
10:000000
```

Fig. 1.10b

Fig. 1.10. (a) and (b) are stem and leaf plots for $\hat{\phi}$ and $\hat{u}=\exp-\hat{\kappa}(1-\cos\hat{\theta})$ for the comet data of Fig. 1.8. (c) is a P-P plot of the $\hat{u}$'s, while (b) and (c) show that there are too many very small and very large $\hat{\theta}$ for the Fisher distribution to be a satisfactory fit. The $\hat{\phi}$'s appear satisfactorily uniform.

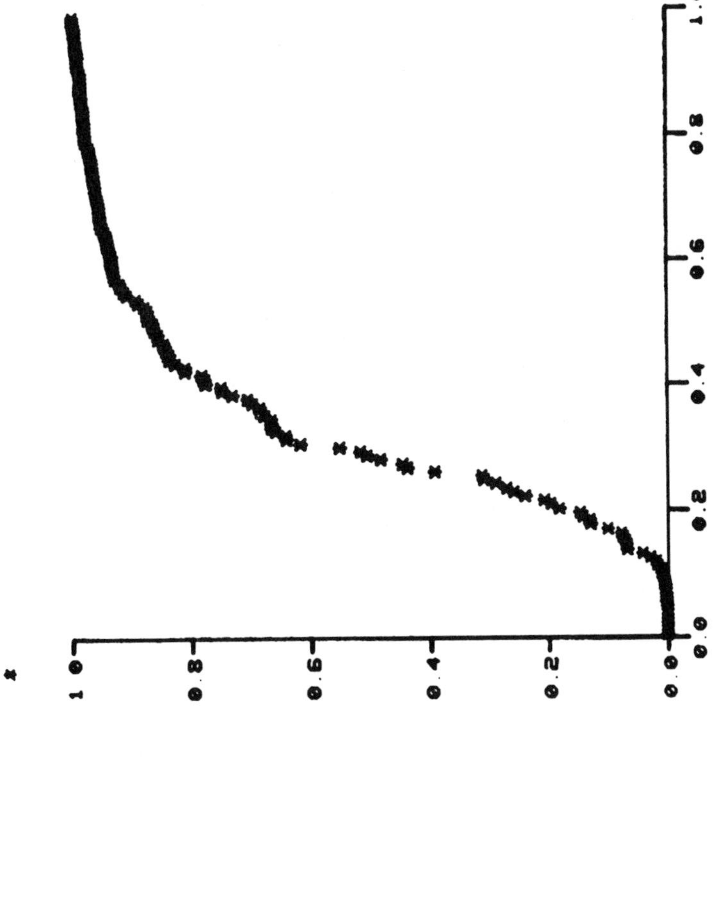

Fig. 1.10c

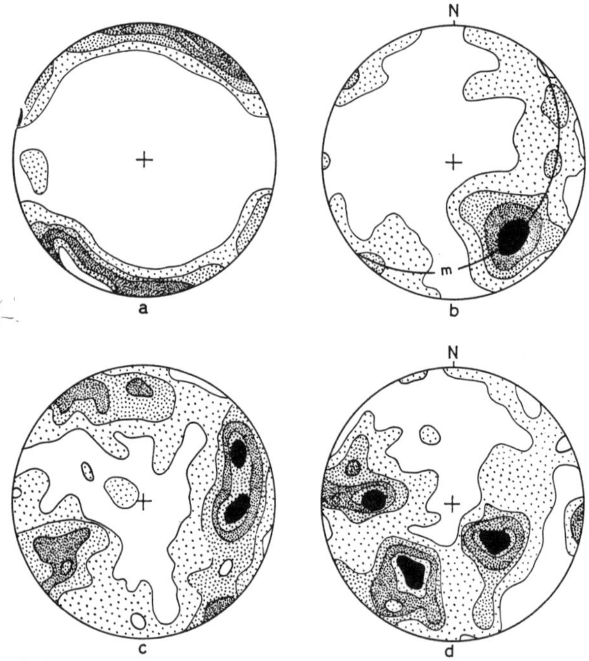

Fig. 1.11. Examples of monoclinic and triclinic subfabrics. (*a*) Monoclinic subfabric; 255 [0001] axes of quartz in mylonitized quartzite from Barstow, California. (After L. E. Weiss.) Contours, 9%, 7%, 5%, 3%, 1%, per 1% area. (*b*) Monoclinic subfabric; 2000 poles to foliation in schists and quartzites from Loch Leven, Scottish Highlands. (After L. E. Weiss.) Contours, 4%, 3%, 2%, 1% per 1% area. (*c*) Triclinic subfabric; 416 [0001] axes of quartz in deformed quartzite pebble from Panamint Range, California. (After L. E. Weiss.) Contours, 5%, 4%, 3%, 2%, 1%, per 1% area. (*d*) Triclinic diagram (heterogeneous body); 193 poles to foliation in gneiss and schist near Lake O'Keefe, East Central Quebec, Canada. (After G. Gastil & L. E. Weiss.) Contours, 5%, 4%, 3%, 2%, 1%, per 1% area. (Taken directly from Turner & Weiss (1963).)

be used. Let $\delta_n(x;z)$ be a probability density on the sphere where x and z lie on the sphere, z fixed, x variable. For example one could choose (1.3.1) and write

$$\delta_n(x;z) = \frac{\alpha_n}{4\pi\sinh\alpha_n} \exp\alpha_n z'x \quad . \tag{1.3.18}$$

With this choice, and denoting the unit sphere in 3 dimensions by Ω_3 and the area element on Ω_3 by $d\omega_3$ (e.g. using spherical polars, $d\omega_3 = \sin\theta d\theta d\phi$) ,

$$\int_{\Omega_3} f(x)\ \delta_n(x;z)d\omega_3 \to f(z) \ , \text{ as } n\to\infty \tag{1.3.19}$$

if $\alpha_n \to \infty$, as $n\to\infty$.

Then consider the estimator $\hat{f}(z)$ of $f(z)$ defined by

$$\hat{f}_n(z) = \frac{1}{n} \sum_{j=1}^{n} \delta_n(x_j;z) \ ,$$

where $x_1,\ldots,x_j$ is a random sample from the density $f(x)$. Then

$$E\hat{f}_n(z) = \int_{\Omega_3} \delta_n(x;z)f(x)dx \ ,$$

$$\to f(z) \ ,$$

if we suppose $\delta_n(x,z)$ is chosen so that (1.3.19) is true. Further,

$$\operatorname{var} \hat{f}_n(z) = \frac{1}{n} \operatorname{var} \delta_n(x;z) \sim \frac{1}{n} [f(z) \int_{\Omega_3} \delta_n^2 (x;z)d\omega_3 - f(z)^2]$$

$$\sim \frac{f(z)}{n} \int_{\Omega_3} \delta_n^2(x;z)d\omega_3 \ , \text{ as } n\to\infty \ .$$

Thus $\operatorname{var} \hat{f}_n(z)$ and the mean square error $E(\hat{f}_n(z)-f(z))^2$ tend to zero as $n\to\infty$, provided $\delta_n(x,z)$ is also chosen so that

$$n^{-1} \int_{\Omega_3} \delta_n^2(x;z)d\omega_3 \to 0 \quad . \tag{1.3.20}$$

This method, using (1.3.18), is easily programmed. The condition (1.3.20) becomes

$$n^{-1}\,\frac{\alpha_n}{8\pi}\,\frac{\sinh 2\alpha_n}{(\sinh\alpha_n)^2} \sim \frac{\alpha_n}{4\pi n} \to 0\ , \tag{1.3.21}$$

which is little help in practice. One may just try a few values of α_n and study the plots. To understand the arithmetic, one should visualize putting Fisher distributions (i.e., little mountains) with total mass $1/n$ over each sample point and then finding the total height at any point z. This was suggested briefly in Watson (1970) and, independently, implemented in the thesis on polar wandering by Van Alstine (1979).

German and McClure (1981) suggested that cross-validation be used to get a more objective method for estimating probability densities. Adapting this idea to our problem, denote the density (1.3.18) with α instead of α_n by $\delta_\alpha(x;z)$. Drop the data point x_k and compute

$$f_\alpha(x_k) = \frac{1}{n-1}\sum_{j\neq k}\delta_\alpha(x_j;x_k) \qquad (k=1,\ldots,n)\ . \tag{1.3.22}$$

A pseudo-likelihood for the data is now

$$\prod_{k=1}^{n} f_\alpha(x_k)\ . \tag{1.3.23}$$

Since (1.3.23) depends upon α, an analogy with maximum likelihood suggests that α be found to maximize (1.3.23) by a numerical maximization process. This method runs into trouble if one is dealing with random variables on say $(0,\infty)$, because, as $n\to\infty$, one tends to get larger and larger observations and these hold down the value of α -- and we know α must tend to infinity as $n\to\infty$. This cannot happen on a bounded interval. Thus we may expect this estimator to be consistent on the sphere. It has given sensible pictures in the trials we have made. Clearly the above methods can be trivially adjusted to work on a sphere in any number of dimensions.

Chapter 2 The Uniform Distribution on Ω_q

2.1 Introduction

Following Müller (1966), we will denote the unit sphere in $\mathbb{R}^q$, $x_1^2+\ldots+x_q^2 = 1$, by Ω_q. On occasions, Ω_2 and Ω_3 will be called the circle and sphere. $U(\Omega_q)$, the uniform distribution on Ω_q, plays a central role both mathematically and scientifically.

In 1734, D. Bernoulli won a prize from the French Academy for an essay on orbits of the planets. Before outlining a theory for why the (then 5 known) planetary orbital planes are similar, Bernoulli attempted to show that this similarity would have been most unlikely to have occurred by chance. Using the right hand rule, each orbit corresponds to a point on the sphere, and he made the first attempt to construct a significance test! In this case, the null hypothesis was that the 5 points were drawn from a uniform distribution on the sphere (Ω_3).

Lord Rayleigh (1880) clarified the question of what is the intensity of the superposition of a large number of vibrations of the same frequency but of arbitrary phase by explicitly assuming that the phases were independently and uniformly distributed on $(0,2\pi)$ and seeking a proper limiting distribution. The superposition could be written $\sum_1^n \exp i(f\theta + \theta_j)$, where the θ_j are the random phases, and it suffices to consider $\sum_1^n \exp i\theta_j$. He identified the $\exp i\theta_j$ with unit vectors and considered first the case where $\text{Prob}(\theta_j=0)=1/2=\text{Prob}(\theta_j=\pi)$ by using the Gaussian approximation to the binomial. Thereafter, for θ_j uniform on $(0,2\pi)$, his argument is a variant of observing that $E(\cos\theta_j)=E(\sin\theta_j) = 0$, $E(\cos^2\theta_j)=E(\sin^2\theta_j) = 1/2$,

$E(\cos\theta_j \sin\theta_j) = 0$,

$$\| \Sigma \exp i\theta_j \|^2 = (\Sigma\cos\theta_j)^2 + (\Sigma\sin\theta_j)^2 \quad ,$$

and that $(2/n)^{1/2}\Sigma\cos\theta_j$, $(2/n)^{1/2}\Sigma\sin\theta_j$ become independent and standard Gaussian as $n\to\infty$ by the Central Limit Theorem. Thus

$$\mathcal{L}\, \frac{2}{n}\{(\Sigma\cos\theta)^2 + (\Sigma\sin\theta)^2\} \to \chi_2^2 \tag{2.1.1}$$

or

$$\mathcal{L}\frac{2}{n}\, \|X\|^2 \to \chi_2^2 \quad ,$$

if we write $x_j = (\cos\theta_j, \sin\theta_j)$, $X = x_1 + \ldots + x_n$. Here the symbol $\mathcal{L}$ is short for "the probability law of" and χ_p^2 is short for "the chi-squared distribution with p degrees of freedom". We may also use it to mean a random variable with this distribution.

Sir Ronald Ross, who won the Nobel Prize in 1902 for his work on Malaria, needed to know how the density of mosquitoes would fall as the distance from their breeding place increased. Ross had always been interested in mathematics -- see his memoirs (1923) -- and founded the mathematical theory of epidemics, partly by his own papers, but more by inspiring A.G. McKendrick who worked for him briefly. Though Ross obtained a solution, he asked Karl Pearson for assistance in 1904. Pearson (1905, July 27) asked readers of Nature if they could solve the problem of random walk in the plane for n steps. Rayleigh (1905, August 3) gave his 1880 asymptotic solution. Kluyver (1906) responded with an elegant study, using an integral representation of a "discontinuous" factor, beginning with unequal steps. He showed that the probability after n unit steps, for a mosquito to be within c of the breeding point is given by

$$c \int_0^\infty J_1(uc)\, J_0(u)^n du \tag{2.1.2}$$

where J_0 and J_1 are Bessel functions. Pearson (1906) numerically evaluated this integral for $n \leqslant 7$ as an aid in evaluating the accuracy of an asymptotic expansion in Laguerre polynomials, valid as $n \to \infty$. Rayleigh (1919) returned to the topic and treated it in great detail. Kluyver's argument and result are clarified and shown to agree with the asymptotic result (2.1.1.) and a similar argument is developed for the three dimensional case. The analogue of (2.1.1.) in q dimensions is trivial to derive by the method used for (2.1.1.) and is

$$\mathcal{L}\, \frac{q}{n} \|X\|^2 \sim \chi_q^2 \;, \tag{2.1.3}$$

where the x_j in $X = x_1 + \ldots + x_n$ are independently and uniformly distributed on Ω_q.

The literature on random walks is now huge and outside our province. However from the last chapter it should be clear that we will need to know the distribution $X = x_1 + \ldots + x_n$. Since here the x_j are uniformly distributed, the direction of X, $X/\|X\|$, must also be uniformly distributed. Thus when we know the distribution of $\|X\|$, we know the distribution of X. Hence these and related results are of interest to us.

The discussion in Section 1.3 of the eigenvalues of the matrix $M_n = n^{-1} \sum_1^n x_j x_j'$ suggests that we should find their joint distribution when the x_j are $U(\Omega_q)$. The matrix M_n is studied in Sections 2.3 and 2.4.

In Section 2.5, the subject of testing for uniformity is briefly described. In Section 2.6, the topic is the computer generation of pseudo uniform vectors on Ω_q . The final section gives a short account of some limiting results when $q \to \infty$.

In this chapter, we will also establish some notation and consider mathematical preliminaries. On several occasions here and later, we will need to call upon a central limit theorem for unit vector random variables. The theorem quoted in Rao (1973) for independent and identically distributed vectors $x \in \mathbb{R}^q$ with $Ex=\mu$, $E(x-\mu)(x-\mu)' = \Sigma$ implies that if $X = x_1 + \ldots + x_n$, the distribution of $\sqrt{n}\,(\frac{X}{n} - \mu)$ tends to the Gaussian distribution in $\mathbb{R}^q$ with a zero mean vector and covariance matrix Σ . We will abbreviate such statements by saying

$$\mathcal{L}\,\sqrt{n}\,(\frac{X}{n} - \mu) \to G_q(0,\Sigma) \quad .$$

2.2 The distribution of $X = x_1 + \ldots + x_n$

In the notation of Müller (1966), let $d\omega_q$ be the area element on $\Omega_q = \{x | x \in \mathbb{R}^q , \|x\|=1\}$ and ω_q be the area of Ω_q. Here, $\|x\|^2 = x_1^2 + \ldots + x_q^2$. If $\varepsilon_1, \ldots, \varepsilon_q$ are orthonormal vectors in $\mathbb{R}^q$, and $x \in \Omega_q$, then

$$x = t\,\varepsilon_q + \sqrt{1-t^2}\,\xi_{q-1} , \quad -1 \leqslant t \leqslant 1 , \tag{2.2.1}$$

where $t = x'\varepsilon_q$, ξ_{q-1} = unit vector in the space spanned by $\varepsilon_1, \ldots, \varepsilon_{q-1}$ and

$$d\omega_q = (1-t^2)^{(q-3)/2} dt\, d\omega_{q-1} . \tag{2.2.2}$$

(The special case $d\omega_3 = d\cos\theta d\phi$ is well known.) Thus

$$\omega_q = \omega_{q-1} \int_{-1}^{1} (1-t^2)^{(q-3)/2} dt = \frac{\pi^{\frac{1}{2}}\Gamma(\frac{q-1}{2})}{\Gamma(q/2)}\, \omega_{q-1} , \tag{2.2.3}$$

so that by recursion,

$$\omega_q = 2\pi^{q/2} / \Gamma(q/2) . \tag{2.2.4}$$

The characteristic function, or Fourier transform, of any density $f(x)$ on Ω_q is given by

$$\phi(v) = E(\exp iv'x) = \int_{\Omega_q} \exp iv'x\, f(x) d\omega_q , \quad v \in \mathbb{R}^q .$$

The fundamental distribution on Ω_q is the <u>uniform distribution</u> with density ω_q^{-1}. If v in (2.2.4) is written as θa, $\|a\|=1$ we may identify a with ε_q in (2.2.1) so that the characteristic function of

$t=a'x$ is given by $\psi(\theta)$ defined by

$$\psi(\theta) = E(\exp i\theta t) = \frac{\omega_{q-1}}{\omega_q} \int_{-1}^{1} e^{i\theta t} (1-t^2)^{(q-3)/2} dt = \psi(\theta) \tag{2.2.5}$$

on using (2.2.2). Formula (9.1.20) in Abramowitz and Stegun (1965) allows us to write

$$E(\exp i\theta t) = \Gamma(q/2)\, J_{q/2-1}(\theta)(\theta/2)^{-q/2+1} , \tag{2.2.6}$$

where $J_\nu(z)$ is a Bessel function of the first kind. Using the modified Bessel function $I_\nu(z)$ ((ibid) formula (9.6.3), $J_\nu(iz) = i^\nu I_\nu(z)$)

$$E(\exp\theta t) = \Gamma(q/2)\, I_{q/2-1}(\theta)(\theta/2)^{q/2-1}$$

is the moment generating function of t .

From (2.2.2), the density of t is immediately seen to be

$$\frac{\omega_{q-1}}{\omega_q} (1-t^2)^{(q-3)/2} . \tag{2.2.7}$$

Thus for $q=2$, the density is U-shaped, flat when $q=3$, and, as q increases, it has an increasing peak at the origin and smaller tails. Hence we may expect that the larger q , the more rapidly the sum of independent t's will become Gaussian.

The characteristic function of

$$y = a'X = a'x_1 +\ldots+ a'x_n = t_1 +\ldots+ t_n$$

is the nth power of (2.2.6). Hence the density of y is given by

Fourier inversion as

$$\frac{1}{2\pi} \int_{-\infty}^{\infty} \exp(-iy\theta)\{\Gamma(q/2)J_{q/2-1}(\theta)(\theta/2)^{-q/2+1}\}^n \, d\theta \, . \qquad (2.2.8)$$

When $q=2$, (2.2.8) reduces, because $J_0(\theta)$ is even in θ , to

$$\frac{1}{\pi} \int_0^{\infty} \cos y \;\; \theta \, J_0(\theta)^n \, d\theta \quad , \qquad (2.2.9)$$

as given by Lord (1948), and distribution function

$$\frac{1}{2} + \frac{1}{\pi} \int_0^{\infty} \frac{\sin y\theta}{\theta} J_0(\theta)^n \, d\theta \quad . \qquad (2.2.10)$$

If $r = \|X\|$, $y = r\cos\phi$ where ϕ , the angle between a and X , is uniform on $(0,2\pi)$ and independent of r . Since $\text{Prob}(r < c)=\text{Prob}(y < c \cos\phi)$, we find from (2.2.9) and the integral equation (2.2.15) below, that

$$\text{Prob}(r < c) = c \int_0^{\infty} J_1(c\theta)J_0(\theta)^n \, d\theta \quad , \qquad (2.2.11)$$

Kluyver's formula, after a short calculation using the facts that

$$J_0'(z) = -J_1(z) \quad , \; J_0(z) = \frac{2}{\kappa} \int_0^1 \frac{\cos zt dt}{(1-t^2)^{\frac{1}{2}}} \quad . \qquad (2.2.12)$$

Here (2.2.15) can be written

$$yu_n(y) = \frac{y}{\pi} \int_y^n \frac{g(r)dr}{(r^2-y^2)^{\frac{1}{2}}} \quad ,$$

and can be solved by multiplying both sides by $(y^2-z^2)^{-\frac{1}{2}}$, integrating the r.h.s. with the substitution $y^2 = r^2\cos^2\theta + z^2\sin^2\theta$.

Durand and Greenwood (1955,1957) discussed this problem in detail and tabulated $\text{Prob}(r < c)$ for various n and c. Statisticians have prepared inverse tables -- Stephens (1969), Batschelet (1971), Mardia (1972)-- to use for the significance test, mentioned in Section 1.2, of the null hypothesis of uniformity. For practical purposes (e.g., $n > 5$), this test can be based on the approximation (2.1.1). We will question the null hypothesis (with a unimodal alternative in mind) if the observed $\|X\|^2$ exceeds $(n/2)$ times 5.99, the 95% point of the χ_2^2 distribution.

When $q=3$, (2.2.5) reduces to

$$\frac{1}{2}\int_{-1}^{1} e^{i\theta t}\, dt ,$$

so that t is seen to be uniformly distributed on $(-1,1)$. This is a consequence of the fact that, if a sphere is circumscribed by a cylinder whose generators are parallel to a, there are equal areas on the cylinder and sphere between two planes with normals parallel to the generators. Thus y is the sum of n independent variables, each uniform on $(-1,1)$, and this distribution appears often and is well studied -- see e.g., Fisher (1953), Feller (1971, p.29). The density and distribution functions of y are given by

$$u_n(y) = \frac{1}{2^n(n-1)!}\sum_{j=o}^{m} (-1)^j\binom{n}{j}(x+n-2j)_+^{n-1} ,$$

$$U_n(y) = \frac{1}{2^n n!}\sum_{j=o}^{m} (-1)^j\binom{n}{j}(x+n-2j)_+^{n} , \qquad (2.2.13)$$

where $(z)_+ = 0$ if $z < 0$, $= z$ if $z \geqslant 0$.

Using again the fact that the direction of X will be uniformly distributed (on Ω_3) and independent of $r = \|X\|$, it is easily shown that the density of r , $v_n(r)$ is given by

$$v_n(r) = -2r\, u_n'(r) \; . \tag{2.2.14}$$

To prove this and other results, note that the density functions of $\cos\phi$ and r are $h(\cos\phi)$ and $g(r)$, and that $y = r\cos\phi$. Then, if $y > 0$, the independence of r and $\cos\phi$ gives

$$u_n(y)dy = \iint_{(y \leqslant r\cos\phi \leqslant y+dy)} g(r)h(\cos\phi)dr\, d\cos\phi$$

$$= \int_y^n g(r)dr \int_{r^{-1}y}^{r^{-1}(y+dy)} h(\cos\phi)d\cos\phi \; ,$$

therefore

$$u_n(y) = \int_y^n g(r)h(\tfrac{y}{r})\, \frac{dr}{r} \; . \tag{2.2.15}$$

Here $h(\cos\phi) = 1/2$ so that

$$u_n(y) = \int_y^n g(r)\, \frac{dr}{2r} \; ,$$

whose derivative is (2.2.14) if we set $g(r) = v_n(r)$.

Again there have been many studies and tabulations (e.g., Watson (1956), Vincenz and Bruckshaw (1957), Stephens (1964)) of this distribution of $\|X\|$ and its inverse function. For testing randomness, (2.1.3) will usually be accurate enough. The 95% point of χ_3^2 is 7.81.

Turning to general q , (2.2.15) is then difficult to solve. However, since $y = rt$, r and t independent,

$$E(\exp i\Theta y) = E_r E_t(\exp i\Theta rt) ,$$

so (2.2.16)

$$\psi^n(\theta) = \int_0^n g(r)\psi(\Theta r)dr$$

where $\psi(\theta)$, by (2.2.6), is given by

$$\psi(\theta) = \Gamma(q/2)\, J_{q/2-1}(\theta)(\theta/2)^{-q/2+1} . \tag{2.2.17}$$

G.N. Watson (1941, p456) shows that if a function $f(r)$ is such that $\int_0^\infty r^{\frac{1}{2}}|f(r)|\, dr < \infty$, and

$$h(\theta) = \int_0^\infty uf(u)J_s(\theta u)du , \tag{2.2.18}$$

then for any $s \geqslant -\frac{1}{2}$ and $r > 0$, (2.2.19)

$$f(r) = \int_0^\infty \theta\, J_s(\theta r)h(\theta)d\theta$$

Suppose that $s=(z/2)-1$. Then (2.2.16) may be rewritten as

$$\Gamma(q/2)^{n-1}J_s(\theta)^n(\theta/2)^{-s(n-1)} = \int_0^\infty r^{-s}g(r)\, J_s(\theta r)dr . \tag{2.2.20}$$

Comparison with (2.2.18) shows that we should set

$$h(\theta) = \Gamma(q/2)^{n-1}\, J_s(\theta)^n(\theta/2)^{-s(n-1)} , \quad f(r) = r^{-s-1}g(r) .$$

Formally (2.2.19) yields

$$r^{-(s+1)}g(r) = 2^{s(n-1)}\Gamma(q/2)^{n-1} \int_0^\infty \Theta^{-sn+s+1} J_s(\theta r)\, J_s(\theta)^n \, d\Theta \,. \tag{2.2.21}$$

Observe that when $q=2$, $s=0$ and (2.2.21) becomes

$$g(r) = r \int_0^\infty \theta\, J_s(\theta r) J_s(\theta)^n \, d\theta \,. \tag{2.2.22}$$

Since $d/dx(xJ_1(x)) = xJ_0(x)$, integration of (2.2.22) from 0 to r yields (2.1.12) for the cumulative distribution function of r.
$\int_0^\infty r^{\frac{1}{2}}|f(r)|dr = \int_0^n r^{-s+\frac{1}{2}}g(r)dr$, which will be bounded if $g(r)$ goes to zero fast enough as r tends to zero. This may be shown to be so by continued differentiation of (2.2.15), provided the required derivative of $u_n(y)$ exists at $y=0$.

G.N. Watson (1944, p.221) extends Kluyver's method to q dimensions by using the discontinuous factor

$$r^{q/2} \int_0^\infty J_{q/2}(r\Theta)\, J_{q/2}(x\Theta)\, \Theta^{-q/2+1}\, d\theta = \begin{cases} 1\,, & x<r \\ 0\,, & x>r\,, \end{cases}$$

and finds the cumulative distribution of $r = \|X\|$ to be

$$r\Gamma(q/2)^{n-1} \int_0^\infty (\tfrac{1}{2}\, r\theta)^{q/2-1}\, J_{q/2}(r\theta)\, \left(\frac{J_{q/2-1}(\theta)}{\frac{1}{2}\theta} \right)^n d\theta \tag{2.2.23}$$

for any integer q. It is easily verified that (2.2.21) is the derivative of (2.2.23).

When n is large, it will be possible to approximate the distribution of $X = x_1 + \ldots + x_n$ by using the Central Limit Theorem (C.L.T.) mentioned in Section 2.1, so that we need to know $\mu = Ex$, $\Sigma = E(x-\mu)(x-\mu)'$ when $\mathcal{L}x$ is $U(\Omega_q)$. If H is an orthogonal matrix, $\mathcal{L}Hx = U(\Omega_q)$ so $\mu = H\mu$, $\Sigma = H\Sigma H'$ for all such matrices. Hence μ is the zero vector and Σ proportional to the identity matrix I_q . Since trace Σ = trace Exx' = E trace $x'x = 1$, we have proved that $\Sigma = I_q/q$. Thus by the C.L.T.,

$$\mathcal{L}n^{-\frac{1}{2}} X \to G(0, I_q/q) \tag{2.2.24}$$

or

$$\mathcal{L}(q/n)^{\frac{1}{2}} X \to G(0, I_q) \quad .$$

An immediate consequence is (2.1.3),

$$\mathcal{L}\frac{q}{n} \|X\|^2 \to \chi_q^2 \quad ,$$

as asserted earlier.

We may expect (2.2.24) to be accurate for quite small n . For $q=1$, when $x=\pm 1$ with probability -½, this is a well known fact. The general theorems in Bhattacharya and Rao (1976) on the rate of convergence of $n^{-\frac{1}{2}}(x_1+\ldots+x_n)$ do not easily yield helpful results. However for every unit vector a , $z=(q/n)^{\frac{1}{2}}a'X$ has the same distribution, and from (2.2.3) it should tend to $G_1(0,1)$. From (2.2.8), the density of z , $f_n(z)$ say, is given by

$$\frac{1}{2\pi} \int_{-\infty}^{\infty} \exp(-i(\tfrac{n}{q})^{\frac{1}{2}} z\theta)\{\Gamma(q/2)J_{q/2-1}(\theta)(\theta/2)^{-q/2+1}\}^n d\theta(\tfrac{n}{q})^{\frac{1}{2}} . \qquad (2.2.25)$$

Setting $\nu=q/2-1$, the expression in braces is (see Abramowitz and Stegun formula (9.1.10))

$$\sum_{k=0}^{\infty} \frac{\Gamma(\nu+1)}{\Gamma(k+\nu+1)} \frac{(\frac{1}{4}\theta^2)^k}{k!} = 1 - \frac{\theta^2}{4(\nu+1)} + \frac{1}{32}\frac{\theta^4}{(\nu+1)(\nu+2)} + O(\theta^6) , \qquad (2.2.26)$$

so the maximum near the origin is at $\theta=0$. Setting $\phi = (\frac{n}{q})^{\frac{1}{2}}\theta$, the density of z is

$$\frac{1}{2\pi} \int_{-\infty}^{\infty} \exp(-iz\phi)\{1 - \frac{\phi^2}{2n} + \frac{q}{q+2}\frac{\phi^4}{8n^2} - \ldots\}^n d\phi$$

$$= \frac{1}{2\pi} \int_{-\infty}^{\infty} \exp(iz\phi - \frac{\phi^2}{2})\{1 - \frac{\phi^4}{4(q+2)n} + O(\frac{1}{n^2})\ d\phi ,$$

$$= \frac{1}{\sqrt{2\pi}} e^{-z^2/2} - \frac{1}{4(q+2)n}(\frac{d}{dz})^4 \frac{1}{\sqrt{2\pi}} e^{-z^2/2} + O(\frac{1}{n^2})$$

$$= \frac{1}{\sqrt{2\pi}} e^{-z^2/2} (1 - \frac{1}{4(q+2)n} H_4(z) + O(\frac{1}{n^2})) , \qquad (2.2.27)$$

where $H_4(z)$ is the fourth order Hermite Polynomial and equal to $z^4 - 6z^2 + 3$. The correction term to the $G_1(0,1)$ limiting distribution does, as predicted, decrease as q increases and its integral from $-\infty$ to ζ is given by

$$\frac{1}{4(q+2)n} \frac{1}{\sqrt{2\pi}} (\zeta^3 - 3\zeta)\exp(-\zeta^2/2) . \qquad (2.2.28)$$

This has its extremes when $\zeta^2 = (3 \pm \sqrt{6})$. The maxima are at

$\zeta^2 = (3 - \sqrt{6}) = 0.551$, so $\zeta = \pm.742$. Hence we find using only the $O(n^{-1})$ correction, that the difference between the true and Gaussian approximation distribution functions cannot exceed

$$\frac{.138}{n(q+2)} .$$

Thus if $q=3$, this bound is .03/n, so already for $n=10$ it is .003, a bearable error in significance tests.

(2.2.27) may be used to find an asymptotic expansion for X if we set $\Theta = \theta a$. For then it follows that

$$E(\exp i(q/n)^{\frac{1}{2}} \Theta' X) = \exp(-\|\Theta\|^2/2)(1 - \frac{1}{4(q+2)n} \|\Theta\|^4) , \tag{2.2.29}$$

the characteristic function of a symmetric q-dimensional distribution. To obtain the first term for the expansion of the density of X, we begin with the known identity,

$$(2\pi)^{-q} \int \exp(iY'\theta - \|\theta\|^2/2)d\theta = (2\pi)^{-q/2}\exp(-\|Y\|^2/2) . \tag{2.2.30}$$

By a suitable set of differentiations of (2.2.30), it follows that

$$(2\pi)^{-q} \int \|\theta\|^4 \exp(iY'\theta - \|\theta\|^2/2)d\theta$$
$$= (2\pi)^{-q/2}\{3-6\|X\|^2 + \|X\|^4\} \exp(-\|Y\|^2/2) . \tag{2.2.31}$$

It will be noted that (2.2.31) agrees with the result implied in (2.2.27) when $q=1$. Thus inverting the charactistic function (2.2.29) shows that the required expansion for the density of $Y = (q/n)^{\frac{1}{2}}X$ is given by

$$\frac{1}{(2\pi)^{q/2}} \exp(-\|Y\|^2/2) \{1 - \frac{1}{4(q+2)n} H_4(\|Y\|)\} . \tag{2.2.32}$$

This density is spherically symmetrical since it depends only on $\|Y\|$.

To obtain the corresponding expansion of the distribution of

$$\|Y\|^2 = \frac{q}{n}\|X\|^2 ,$$

we observe the density $\|Y\|$ is clearly given by

$$\frac{\omega_{q-1}}{(2\pi)^{q/2}} \exp(-\|Y\|^{2/2})\{1 - \frac{1}{4(q+2)n} H_4(\|Y\|)\} \|Y\|^{q-1} \tag{2.2.33}$$

by integrating (2.2.32) for constant $\|Y\|$. The leading term of (2.2.33) simply gives another proof that

$$L\,\frac{q}{n}\|X\|^2 \to \chi_q^2 \quad , n \to \infty .$$

The $O(n^{-1})$ gives a correction to this result, but we cannot pause to discuss it further.

2.3 The distribution of $M_n = n^{-1}\sum_1^n x_i x_i'$

As pointed out in Section 2.1, M_n is a symmetric non-negative matrix with unit trace. The expectation of M_n, since $Ex_i=0$, is the covariance matrix of x which was shown to be I_q/q. Thus by the weak Law of Large Numbers (L.L.N.), $M_n \to I_q/q$ in prob. as $n\to\infty$ but more is true. If $x_i = (x_{i1},\ldots,x_{iq})$, then the (j,k) element of M_n is $n^{-1}\sum_{i=1}^n x_{ij}x_{ik}$, and since $x\in\Omega_q$, $E(x_{ij}x_{ik})^2<\infty$, the C.L.T. will ensure that each element becomes Gaussian. In fact, the $q(q+1)/2$ functionally independent elements of M_n become jointly Gaussian for the same reason. There seems to be no hope of finding their joint distribution for finite n, so we seek their asymptotic distribution. To do so, we must find the expectations of all fourth order products of the components of x.

For the moment, write $x = (x_1,\ldots,x_q)$ so that we need $E(x_i x_j x_k x_\ell)$ for all $i,j,k,\ell=1,\ldots,q$, which can be found in several ways.

Anderson and Stephens (1972) use the fact that if y is a Gaussian random vector with zero mean vector and covariance matrix I_q, i.e., $\mathcal{L}(y) = G_q(0,I_q)$, and $z = y/\|y\|$, then $\mathcal{L}z = U(\Omega_q)$, z and $\|y\|$ are independent, and $\mathcal{L}(\|y\|^2) = \chi_q^2$. Denote $zz' = xx'$ by $A = \{a_{k\ell}\}$. Then $E(A) = Eyy'/E\|y\|^2 = I_q/q$, as before. But

$$E\,a_{kk}^2 = E\,z_k^4 = E\,y_k^4/E\|y\|^4 = 3/q(q+2)\ ,$$

$$E\,a_{k\ell}^2 = Ea_{kk}a_{\ell\ell} = E\,z_k^2z_\ell^2\ ,$$

$$= E\,y_k^2y_\ell^2/E\|y\|^4 = \frac{1}{q(q+2)}\ ,\qquad (k\neq\ell)$$

$$E\,a_{kk}a_{k\ell} = E\,z_k^3z_\ell = Ey_k^3y_\ell/E\|y\|^4 = 0 \qquad (k\neq\ell)\ ,$$

on using normal and chi-square moments. The variances and covariances of the elements of A are therefore

$$\text{var } a_{kk} = \frac{2(q-1)}{q^2(q+2)} , \qquad \text{var } a_{k\ell} = \frac{1}{q/(q+2)},$$

$$\text{cov }(a_{kk}, a_{\ell\ell}) = \frac{-2}{q^2(q+2)} , \text{ other covariances} = 0 . \tag{2.3.1}$$

If now the $q(q+1)/2$ functionally independent terms of A are arranged in a vector as follows, this and its covariance matrix may be displayed informally as

$a_{11} \cdots a_{qq}$	$a_{12} \cdots a_{1q}$	$a_{23} \cdots a_{2q}$	$\cdots$ a_{q-1q}
$\frac{2}{q(q+2)}(I_q - \frac{11'}{q})$	0	0	0
0	$\frac{1}{q(q+2)} I_{q-1}$	0	0
0	0	$\frac{1}{q(q+2)} I_{q-2}$	0

(2.3.2)

Finally the CLT tells us that if the functionally independent elements of $n^{\frac{1}{2}}(M_n - I_q/q)$ are arranged in the same way, then their asymptotic distribution is $G_{q(q+1)/2}(0,(2.3.2))$.

To obtain the matrix (2.3.2), we have been able to use a trick which is only available when $Lx = U(\Omega_q)$. It is worthwhile to show here a more formal method that will work in more general cases which we will meet later. Let $A \otimes B = [a_{ij}B]$. The properties of the product are given in MacDuffee (1946) (see also Neudecker (1968), Magnus and Neudecker (1979) for more detailed statistical applications). We will use the facts

$$\begin{aligned} (A \otimes B)' &= A' \otimes B' , \\ (A_1 \otimes B_1)(A_2 \otimes B_2) &= (A_1A_2) \otimes (B_1B_2) , \\ \text{trace } (A \otimes B) &= (\text{trace } A)(\text{trace } B) , \\ xx' &= x \otimes x . \end{aligned} \qquad (2.3.3)$$

If $\mu_1,\ldots,\mu_q$ is an orthonormal basis (ONB) in $\mathbb{R}^q$, then the q^2 vectors $\mu_i \otimes \mu_j$, $i,j = 1,\ldots,q$ form an ONB in $\mathbb{R}^q$ because

$$\begin{aligned} \| \mu_i \otimes \mu_j \|^2 &= (\mu_i' \otimes \mu_j')(\mu_i \otimes \mu_j) \\ &= (\mu_i'\mu_i) \otimes (\mu_j'\mu_j) \\ &= 1 \quad , \text{ for all } i,j , \end{aligned}$$

and

$$\begin{aligned} (\mu_i \otimes \mu_j)' (\mu_k \otimes \mu_l) &= (\mu_i'\mu_k) \otimes (\mu_j'\mu_l) \\ &= 0 \text{ , unless } i = k \text{ , } j = l \end{aligned}$$

The elements in the matrix (2.3.2) are certain elements in the matrix

$$V = E(x \otimes x)(x' \otimes x') = E(xx') \otimes (xx')$$

when $Lx = U(\Omega_q)$, so that $Ex = 0$, $Exx' = I_q/q$. The symmetric matrix V is easy to study. By (2.3.3),

$$\text{trace } V = E(\text{trace } xx')^2 = 1 \,. \tag{2.3.4}$$

The spectral form of V may be obtained by observing that

$$xx' = x \otimes x = \sum_i \sum_j (\mu_i \otimes \mu_j)'(x \otimes x)(\mu_i \otimes \mu_j) \,,$$

$$= \sum_i \sum_j (\mu_i' x)(\mu_j' x)\mu_i \otimes \mu_j \,,$$

so that

$$V = \sum_i \sum_j \sum_k \sum_l E(\mu_i' x \; \mu_j' x \; \mu_k' x \; \mu_l' x)(\mu_i \otimes \mu_j)(\mu_k' \otimes \mu_l') \,.$$

Because $L\,x = U(\Omega_q)$, this reduces to the spectral form

$$V = \sum_i \lambda_4(\mu_i \otimes \mu_i')(\mu_i' \otimes \mu_i) + \sum_{i \neq j} \sum \lambda_{22}(\mu_i \otimes \mu_j)(\mu_i \otimes \mu_j)' \,, \tag{2.3.5}$$

where

$$\left.\begin{aligned} \lambda_4 &= E(\mu_1'x)^4 = Ex_1^4 \ , \\ \lambda_{22} &= E(\mu_1'x)^2(\mu_2'x)^2 = Ex_1^2x_2^2 \ . \end{aligned}\right\} \quad (2.3.6)$$

Because of (2.3.4),

$$q\lambda_4 + q(q-1)\lambda_{22} = 1 \ .$$

A related method of studying V, particularly relevant when the distribution of x is rotationally symmetric about μ_1, say, is to compute the expectation of $xx' \otimes xx'$ directly by writing $x = t\mu_1 + (1-t^2)^{\frac{1}{2}}\xi$, where $\|\xi\| = 1$, $\xi \perp \mu_1$. The results simplify easily because t and ξ are independent.

After this technical aside, we return to a key problem: testing whether a distribution is uniform on Ω_q.

In Section (1.3), a test of uniformity on Ω_3 used the statistic $\frac{15n}{2}\sum_1^3 (\hat{\lambda}_j - 1/3)^2$, where $M_n\hat{\mu}_j = \hat{\lambda}_j\hat{\mu}_j$. We can now prove the general result for Ω_q,

$$\mathcal{L}\, \frac{n(q+2)q}{2} \sum_1^q (\hat{\lambda}_j - 1/q)^2 \rightarrow \chi^2_{q(q+1)/2 - 1} \ , \quad (2.3.7)$$

due to Anderson and Stephens (1972) for general q, and to Bingham (1964) for $q=3$.

Setting

$$T = n^{\frac{1}{2}}(M_n - I_q/q) \quad ,$$

we find

$$\text{trace } T^2 = n\ \Sigma(\hat{\lambda}_j - 1/q)^2 \ . \tag{2.3.8}$$

But the asymptotic distribution of the functionally independent elements $t_{k\ell}(k \leqslant \ell)$ of T is $G_{q(q+1)/2}\ (0,(2.3.2))$ and

$$\text{trace } T^2 = \Sigma t_{kk}^2 + 2 \sum_{k<\ell} t_{k\ell}^2 \ . \tag{2.3.9}$$

From (2.3.2), we see that

$$\mathcal{L}\, 2 \sum_{k<\ell} t_{k\ell}^2 \to \frac{2}{q(q+2)} \chi^2_{q(q-1)/2} \ , \tag{2.3.10}$$

and that

$$\mathcal{L} \sum_1^q t_{kk}^2 \to \frac{2}{q^2(q+2)} \sum_1^q y_k^2 \quad ,$$

where $y_1,\ldots,y_q$ are Gaussian with zero means, variances $q-1$ and covariances -1. Setting $z_k = y_k q^{-\frac{1}{2}}$, the covariance matrix of the $z_1,\ldots,z_q$ has familiar form -- on the diagonal $1-q^{-1}$, off the diagonal $-q^{-1}$. Hence

$$\mathcal{L}\Sigma\ t_{kk}^2 \to \frac{2}{q(q+2)} \chi^2_{q-1} \ . \tag{2.3.11}$$

Thus, adding (2.3.10) and (2.3.10), the result (2.3.7) is established.

2.4 The joint distribution of the eigen values of M_n

Since trace $M_n = \sum_1^q \hat{\lambda}_j = 1$, the joint distribution $0 \leqslant \hat{\lambda}_1 \leqslant \ldots \leqslant \hat{\lambda}_q \leqslant 1$ is singular. Anderson and Stephens (1972) gave the following ingenious argument.

To get an asymptotic distribution, we must consider the matrix $T = n^{\frac{1}{2}}(M_n - I_q/q)$. The eigenvalues of T add to zero, and so have singular joint distribution. In order to deal with densities, they study first the eigenvalues of

$$U = T + vI_q \,, \quad v = G_1(0,\sigma^2). \tag{2.4.1}$$

Then, using (2.3.2),

$$\operatorname{var}(u_{kk}) = \frac{2(q-1)}{q^2(q+2)} + \sigma^2$$

$$\operatorname{cov}(u_{kk},u_{\ell\ell}) = \frac{-2}{q^2(q+2)} + \sigma^2$$

$$\operatorname{var}(u_{k\ell}) = \frac{1}{q(q+2)} \,, \quad \operatorname{cov}(u_{ij},u_{k\ell}) = 0 \,,$$

so that choosing $\sigma^2 = 2/q^2(q+2)$ makes all the covariances zero, and the functionally independent elements of U have, asymptotically, a non-singular Gaussian distribution and

$$\operatorname{var}(u_{kk}) = 2/q(q+2) \,.$$

Thus the quadratic form in the exponent of this distribution is

$$\frac{q(q+2)}{2} \sum_1^q u_{kk}^2 + q(q+2) \sum_{k<\ell} u_{k\ell}^2 = \frac{q(q+2)}{2} \operatorname{trace} U^2 \,,$$

so the joint density of the independent elements of U is

$$\frac{\{q(q+2)\}^{q(q+1)/4}}{2^{q/2}(2\pi)^{q/(q+1)/4}} \exp - \frac{1}{4} q(q+2) \text{ trace } U^2 . \qquad (2.4.2)$$

It then follows from Theorem 13.3.1 in Anderson (1958) that the joint density of the roots $s_1<\dots<s_q$ of U is

$$\frac{\{q(q+2)\}^{q(q+1)/4}}{2^{q(q+3)/4} \prod_{i=1}^{q} \Gamma\{\frac{1}{2}(q-i+1)\}} \exp\{-\frac{q(q+2)}{4} \sum_1^q s_i^2\} \prod_{i>j} (s_i-s_j) . \qquad (2.4.3)$$

The roots of r_i of T and s_i of U are simply related by $s_i = r_i + v$ because of (2.4.1). Trace T = 0 = Σr_i , so $\Sigma s_i = qv$ or $v = \bar{s}$, $r_i = s_i - \bar{s}$,

$$\Sigma s_i^2 = \Sigma r_i^2 + qv^2 .$$

The joint density, with $r_q = -(r_1+\dots+r_{q-1})$, of v and $r_1,\dots,r_{q-1}$ is then

$$\left[\frac{q^{q(q+1)/4}(q+2)^{(q(q+1)/4)-(\frac{1}{2})}}{2^{q(q+3)/4 - 1} \prod_{j=2}^{q} \Gamma(j/2)} \exp\{-\frac{1}{4} q(q+2) \sum_1^q r_i^2\} \prod_{i>j} (r_i-r_j)\right]$$

$$\times \left[\frac{q(q+2)^{\frac{1}{2}}}{2\pi^{\frac{1}{2}}} \exp\{-\frac{q^2(q+2)}{4} v^2\}\right] . \qquad (2.4.4)$$

It is seen that v is independent of $r_1, \ldots, r_{q-1}$, whose density is the first factor in (2.4.4). This first factor is thus the density of the limiting distribution of the $q-1$ smallest eigenvalues of $n^{\frac{1}{2}}(M_n - I_q/q)$.

Anderson and Stephens (1972) go on to consider the 3-dimensional case in detail.

2.5 Testing whether f is uniform

If on the null hypothesis H_0, $f=\omega_q^{-1}$ and on the alternative H_1, $f=f_1$, then the Neyman-Pearson Lemma tells us to reject H_0 in favor of H_1 if $\prod_{j=1}^{n} f_1(x_j)$ is too large. Suppose $f_1 = f(0_q x)$ where 0_q is a rotation matrix. If 0_q is unknown, it is reasonable to demand a test statistic which does not depend upon what 0_q is, i.e., to demand an invariant test. This leads to the test statistic (Lehmann, 1959)

$$\int \prod_{j=1}^{n} f(0_q x_j) d0_q = \underset{0_q}{\text{ave}} \prod f(0_q x_j) \ . \tag{2.5.1}$$

The statistic (2.5.1) may be trivially evaluated when $q=2$ because it will naturally be written

$$\int_0^{2\pi} \prod_1^n f(\theta_j+\phi) d\phi \ . \tag{2.5.2}$$

In the particular case where $f(\theta+\phi) = \exp\kappa\cos(\theta+\phi)/2\pi\, I_0(\kappa)$, (2.5.2) is proportional to $I_0(\kappa R)$, which (since $\kappa > 0$) increases monotonically with R (the length of the sum of the data vectors). Thus we can assert that, in this case, the so-called Rayleigh test -- reject uniformity if R is too large -- is the best invariant test. This test makes intuitive sense whenever the alternative is uni-modal, as we have seen earlier.

If the alternative density is very far from uniformity, it should be easy to design a sensitive test when one has enough data. But if the alternative density is close to uniformity, more care is obviously required. Beran (1968) gave an elegant theory for testing for uniformity on compact homogeneous spaces. To sketch this in our setting, we suppose a sequence of alternatives to uniformity defined by

$$f_\kappa(x) = \omega_q^{-1} + \kappa\{f(0_q x) - \omega_q^{-1}\}\ , \quad \kappa \to 0\ . \tag{2.5.3}$$

The integral of $f_\kappa(x)$ over Ω_q will be unity and $f_\kappa(x)$ will be non-negative for κ small enough if $f(x)$ is bounded on Ω_q. Of course, $f_\kappa(x) \to \omega_q^{-1}$, the uniform density as $\kappa \to 0$.

Setting $\lambda = \kappa\omega_q$, Ave $\Pi f(0_q x_j)$ is proportional to

$$\text{Ave}\ \Pi\ (1 + \lambda\{\omega_q\ f(0_q x_j) - 1\})$$

$$= \text{Ave}\ [1 + \lambda \sum_j \{\omega_q\ f(0_q x_j) - 1\} + \lambda^2 \sum\sum_{j \neq k} (\omega_q f(0_q x_j) - 1)(\omega_q f(0_q x_k - 1)]$$

plus smaller terms. The average of the coefficient of λ is zero and the coefficient of λ^2 may be simplified by the identity $\sum\sum_{j\neq k} a_j a_k = (\Sigma a_j)^2 - \Sigma a_j^2$. Noting that

$$\text{Ave}\ (\omega_q f(0_q x_j) - 1)^2 \quad \text{independent of} \quad x_j\ ,$$

Beran thus shows that the best invariant test for this sequence of local alternative hypotheses is based on large values of

$$\underset{0_q}{\text{Ave}}\ [\sum_j \{\omega_q f(0_q x_j) - 1\}]^2\ . \tag{2.5.4}$$

To complete the test, we need the distribution of (2.5.4) when the data actually come from a uniform distribution. This is naturally calculated by Fourier methods. If we take a circle of unit perimeter and set $f(\theta) = \Sigma c_m \exp i2\pi m\theta$, it is easily shown that

$$\Sigma\{f(\theta_j+\phi)-1\} = \sum_{m\neq 0} c_m \exp i2\pi m\phi \sum_j \exp i2\pi m\theta_j \quad , \tag{2.5.5}$$

$$\int_0^1 [\Sigma\{f(\theta_j+\phi)-1\}]^2 d\phi = 2\sum_1^\infty |c_m|^2 |\sum_j \exp i2\pi m\theta_j|^2 \quad . \tag{2.5.6}$$

$$= 2n \sum_1^\infty |c_m|^2 \; |\hat{c}_m|^2 \; .$$

Now it may be shown that, as in the derivation of (2.1.1),

$$|\hat{c}_m|^2 = \frac{2}{n} |\sum_j \exp i2\pi m\theta_j|^2 \to \chi_2^2 \quad , \quad m=1,2,\ldots, \quad (n \to \infty) \; , \tag{2.5.7}$$

and that these random variables are independent. Hence the asymptotic distribution of

$$\frac{1}{n}\int_0^1 [\Sigma\{f(\theta_j+\phi)-1\}]^2 d\phi \tag{2.5.8}$$

is that of

$$\sum_1^\infty |c_m|^2 \; \chi_2^2 \quad .$$

The statistic (2.5.6) has an intuitive interpretation. We will not pause to give this or to explain how the distribution of (2.5.9) may be obtained. The U_n^2 statistic of Watson (1961) is a special case of (2.5.9) and is a circular variant of the Cramer-von Mises statistic.

Beran's work was motivated by Ajne (1968), who defined special sequences of local and distant alternatives, and Watson's use of Fourier methods. It will be noted that to get statistics of the Kolmogorov type, which use the supremum, it is necessary to use distant alternatives.

This was explored further in Watson (1974), in the E.J.G. Pitman Festschrift.

An important point to notice here is that sample distribution functions will not usually arise -- they are natural only on the line. Further, if the circle is a guide, supremum type tests can only be justified on the rather absurd grounds that the alternative to uniformity has a large jump somewhere in its distribution function. However their mathematical interest has led to an enormous literature.

The topic of testing uniformity is one of great mathematical interest since it may be treated in greater generality than many statistical problems. The papers of Gine (1975), Prentice (1978), give a flavor of this work. Wellner (1979) attacks the problem of testing the identity of two spherical distributions, not necessarily uniform.

2.6 Computer simulation of $U(\Omega_q)$

The generation of pseudo-random vectors on Ω_q will be important for checking various mathematical approximations by computer simulation. Naturally $U(\Omega_q)$ is the basic distribution. Some nonuniform distribution methods are suggested in Section 3.7.

Given the prevalence of good subroutines for generating pseudo-standard Gaussians, $G_1(0,1) = \mathcal{L}(z)$, the simplest method is undoubtedly to generate independent $z_1,\dots,z_q$ and set $\|z\| = (z_1^2 + \dots + z_q^2)^{\frac{1}{2}}$, $x = (x_1,\dots,x_q)$, where $x_j = z_j/\|z\|$.

The primitive pseudo-random variable is the uniform distribution on the unit interval. Thus, possibly faster methods will begin with these. A paper giving the earlier references and doing just this, is that of Tashiro (1977). Methods with rejection often become extremely inefficient when q is large, so then direct methods are far preferable.

Clearly with $q=2$, $x = (\cos\theta\ ,\ \sin\theta)$ where $\theta = 2\pi u$, u uniform on (0,1) will be used. When $q=3$, $x = (\sin\theta\cos\phi\ ,\ \sin\theta\sin\phi\ ,\ \cos\theta)$ where ϕ is uniform on $(0,2\pi)$ and $\cos\theta$ is uniform on $(-1,1)$, and $\sin\theta = (1-\cos^2\theta)^{\frac{1}{2}}$ will be used. The direct extension of these methods leads to difficulties.

Sibuya (1962) suggested the following method. Suppose $q=2m$. Order $m-1$ pseudo uniforms on (0,1) and call them $U_1,\dots,U_{m-1}$. Thus

$$0 = U_0 < U_1 < \dots < U_{m-1} < U_m = 1 . \qquad (2.6.1)$$

Set

$$Y_i = U_i - U_{i-1} \ , \ i = 1,\ldots,m \ , \tag{2.6.2}$$

and let $S_1,\ldots,S_m$ be other pseudo-uniforms. Then $x=(x_1,\ldots,x_{2m})$ are defined as $(i=1,\ldots,m)$

$$\left.\begin{aligned} x_{2i-1} &= Y_i^{\frac{1}{2}} \cos 2\pi S_i \ , \\ x_{2i} &= Y_i^{\frac{1}{2}} \sin 2\pi S_i \ . \end{aligned}\right\} \tag{2.6.3}$$

It is obvious that $x \varepsilon \Omega_q$, but less so that $\mathcal{L}x = U(\Omega_q)$. The only point requiring proof is that if $\mathcal{L}x = U(\Omega_q)$, then ${x_{2i}}^2 + {x_{2i-1}}^2 = Y_i$ are distributed as implied by (2.6.2). But we saw earlier that $(i = 1,\ldots,m)$

$${x_{2i}}^2 + {x_{2i-1}}^2 = ({Z_{2i}}^2 + {Z_{2i-1}}^2) \ / \ \sum_1^{2m} Z_i^2 \ , \tag{2.6.4}$$

where the Z_i's are independent standard Gaussians. The fact that the random variables in (2.6.4) are distributed like the gaps (2.6.2) is a well known result in Time Series Analysis. To avoid the computation of the trigonometric functions in (2.6.3), $(\cos 2\pi S \ , \ \sin 2\pi S)$ may be found from two independent uniforms T_1 and T_2 as follows: accept the pair

$$\left(\frac{T_1^2 - T_2^2}{T_1^2 + T_2^2} \ , \pm \ \frac{2T_1T_2}{T_1^2 + T_2} \right) \ \text{if} \ T_1^2 + T_2^2 \leqslant 1 \ ,$$

and assign the signs at random. An accepted pair is identified with $(\cos 2\pi S, \sin 2\pi S)$.

When $q=2m-1$, the preceding procedure may be used to give $(y_1,\ldots,y_{2m})$ on Ω_{2m}. Then $x\varepsilon\Omega_q$ is a normalization of $y_1,\ldots,y_{2m-1}$. He also gives another method.

Tashiro (1977) takes the matter a little further, particularly in considering the efficient generation of ordered uniforms.

2.7 High dimensional spheres

In the previous sections, we have considered $x_1,\ldots,x_n$ independent identically and uniformly distributed on Ω_q for fixed q. Here we will see what happens when n is fixed but $q \to \infty$.

From (2.2.2), we see that

$$a_q(\kappa) = \int_{\Omega_q} \exp(\kappa\mu' x)\omega_q(dx) ,$$

$$= \omega_{q-1} \int_{-1}^{1} \exp(\kappa t)\ (1-t^2)^{\frac{q-3}{2}} dt .$$

thus

$$a_q(q^{\frac{1}{2}}\kappa) = \frac{\omega_{q-1}}{q^{\frac{1}{2}}} \int_{-q^{\frac{1}{2}}}^{q^{\frac{1}{2}}} \exp(\kappa u)(1 - \frac{u^2}{q})^{\frac{q-3}{2}} du ,$$

$$\sim \frac{\omega_{q-1}}{q^{\frac{1}{2}}} (2\pi)^{\frac{1}{2}} \exp(\kappa^2/2) , \quad q \to \infty . \qquad (2.7.1)$$

By Stirling's formula and the formula for ω_{q-1},

$$a_q(a^{\frac{1}{2}}\kappa) \sim 2\pi^{\frac{q-1}{2}} (\frac{q-1}{2})^{-(\frac{q-2}{2})} \exp\ (\frac{q-1}{2} + \tfrac{1}{2}\kappa^2) . \qquad (2.7.2)$$

Now if $Lx = U(\Omega_q)$, the moment generating function (m.g.f.) of $q^{\frac{1}{2}}\lambda' x$ for $\lambda \varepsilon \mathbb{R}^q$, $\|\lambda\| = 1$, is

$$E(\exp\ \theta q^{\frac{1}{2}}\lambda' x) = \int_{\Omega_q} \exp(\theta q^{\frac{1}{2}}\lambda' x)\ \frac{\omega_q(dx)}{\omega_q} ,$$

$$= a_q(q^{\frac{1}{2}}\theta)/\omega_q \qquad (2.7.3)$$

by the previous paragraph. To get the limit of this as $q \to \infty$, we need another consequence of Stirling's formula:

$$\frac{\Gamma(z+a)}{\Gamma(z+b)} \sim z^{a-b}, \quad z \to \infty . \tag{2.7.4}$$

Combining (2.7.3),(2.7.4) with (2.7.1), we find

$$E \exp \theta q^{\frac{1}{2}} \lambda' x \to \exp(\theta^2/2) , \tag{2.7.5}$$

for any unit vector λ. Thus $q^{\frac{1}{2}}\lambda' x$ is asymptotically standard Gaussian.

If λ has zero coordinates except in the first p places, we may write $\theta\lambda = \phi$, and then (2.7.5) reads

$$E \exp q^{\frac{1}{2}} \phi' x \to \exp (\phi'\phi/2) , \tag{2.7.6}$$

which shows that the joint distribution of the first p coordinates of $q^{\frac{1}{2}}x$ is, as $q \to \infty$, $G_p(0,I_p)$, where 0 is the zero vector in $\mathbb{R}^p$. This is Theorem 1 of Stam (1982).

Suppose $x_1, x_2, \ldots, x_n$ (n fixed) are i.i.d $U(\Omega_q)$. Then Stam's Theorem 4 says that the joint distribution of the $N=n(n-1)/2$ scalar products $q^{\frac{1}{2}}x_i x_j (i=j, i,j=1,\ldots,n)$ is asymptotically standard Gaussian in $\mathbb{R}^N$.

Firstly, by the above,

$$E \exp \theta q^{\frac{1}{2}} x_1' x_2 = \frac{1}{\omega_q^2} \int\int \exp q^{\frac{1}{2}} \theta x_1' x_2 \omega_q(dx_1)\omega_q(dx_2)$$

$$= \frac{1}{\omega_q^2} \int a_q(q^{\frac{1}{2}}|\theta|)\omega_q(dx_2) \ ,$$

$$= \frac{a_q(a^{\frac{1}{2}}|\theta|)}{\omega_q} \ ,$$

$$\sim \exp(\theta^2/2) \ ,$$

so the result is true for $n=2$. For $n=3$, the moment generating function

$$E \exp q^{\frac{1}{2}}(\theta_{12}x_1'x_2 + \theta_{13}x_1'x_3 + \theta_{23}x_2'x_3) ,$$

is

$$\omega_q^{-3} \int\int\int \exp q^{\frac{1}{2}}\theta_{12}x_1'x_2 \exp q^{\frac{1}{2}}(\theta_{13}x_1 + \theta_{23} \ x_2)'x_3 \ \omega_q(dx_1)\omega_q(dx_2)\omega_q(dx_3).$$

Intergrating out x_3 , this becomes

$$\omega_q^{-3} \int\int \exp q^{\frac{1}{2}}\theta_{12}x_1'x_2 \ a_q(a^{\frac{1}{2}} \| \theta_{13}x_1 + \theta_{23}x_2 \|) \ \omega_q(dx_1)\omega_q(dx_2)) \qquad (2.7.7)$$

Using (2.7.1) in (2.7.7), it reduces to

$$\frac{(\omega_{q-1}q^{-\frac{1}{2}}(2\pi)^{\frac{1}{2}})}{\omega_q^3} \exp\tfrac{1}{2}(\theta_{13}^2 + \theta_{23}^2) \text{ times}$$

$$\int\int \exp(q^{\frac{1}{2}}\theta_{12}x_1'x_2 + \theta_{13}\theta_{23}x_1'x_2)\omega_q(dx_1)\omega_q(dx_2) .$$

Using (2.7.1) again on integrating out x_2 , and then integrating out x_1 , we get

$$\frac{(\omega_{q-1}{}^{\frac{1}{2}}(2\pi)^{\frac{1}{2}})^2}{\omega_q^3} \quad \omega_q \exp\tfrac{1}{2}(\theta_{13}^2 + \theta_{23}^2 + \theta_{12}^2) ,$$

which is asymptotically

$$\exp\tfrac{1}{2}(\theta_{12}^2 + \theta_{13}^2 + \theta_{23}^2) ,$$

so the result is true for $n=3$. It is clearly true for any n .

The length $\|X\|$ of $X=x_1+\ldots+x_2$ was of interest above. Now

$$\|X\|^2 = n + 2 \sum_{i<j} x_i x_j ,$$

so by Stam's Theorem 4 we can get the asymptotic distribution of $\|X\|$. Consider

$$\frac{q^{\frac{1}{2}}}{2} (\|X\|^2 - n) = q^{\frac{1}{2}} \sum_{i<j} x_i' x_j . \qquad (2.7.8)$$

Since the r.h.s of (2.7.8) has, as $q \to \infty$, a $G_1(0,N)$ distribution, we see that, as $q \to \infty$

$$\|X\| \to n^{\frac{1}{2}} \text{ (in prob.) } . \qquad (2.7.9)$$

Combining (2.7.8) and (2.7.9), we have proved that

$$L(nq^{\frac{1}{2}})(\|X\| - n^{\frac{1}{2}}) \to G_1(0,N) \ . \qquad (2.7.10)$$

We will show elsewhere that all these results are easily generalized to non-uniform distributions. The asymptotic discussion of the eigenvalues of M_n considered in Section (2.4), as $q \to \infty$, leads into deeper waters, and it too will be discussed elsewhere.

Chapter 3 Classes of Distributions on Ω_q

3.1 Introduction

Any non-negative measure $\mu(\cdot)$ on Ω_q such that $\mu(\Omega_q) = 1$ suffices to define a probability-distribution on Ω_q. Statisticians will however tend to use classes of distributions which

(i) fit the data at hand,

(ii) are relatively easy to handle mathematically and statistically,

(iii) arise from stochastic models of relevant processes.

These criteria are rather vague. The examples in Chapter 1 have given some typical data sets and some distributions have been suggested, so this chapter will concentrate on some implications of (ii) and (iii). These topics are obviously *very* broad -- think of the vast literature devoted to them for real valued random variables, much of it very theoretical. Even the choice of a few subtopics that seem to throw light on practical statistics is rather subjective. Further, the subject is underdeveloped so that we cannot give references to full accounts of matters ignored here. We will be satisfied if readers find this chapter interesting but incomplete. A recurring theme will be analogies with the Gaussian distribution, which is so easy to deal with.

The likelihood of a sample $x_1, \ldots, x_n$ is $\prod_1^n f(x_i,\alpha)$ if it is drawn from a distribution with density $f(x,\alpha)$ (with respect to the invariant measure $\omega(\cdot)$ of Chapter 2), where α is a parameter. Thus it will be convenient if f is proportional to the exponential of a simple function, e.g., one that is linear in the components of $\boldsymbol{\alpha}$ (canonical form), or in functions of them. This leads to *exponential families* -- see, e.g., Barndorff-Nielsen

(1978) for a detailed study. The Gaussian distribution on $\mathbb{R}$ belongs to this family, as do the von Mises and Fisher distributions on the circle and the sphere, or more generally the Langevin distribution on Ω_q, whose density is $\kappa \geq 0_1\mu$, $x \in \Omega_q$

$$(\text{functions of } \kappa) \exp \kappa\, \mu' x = (\text{functions of} \|\lambda\|) \exp \lambda' x , \qquad (3.1.1)$$

where $\lambda = \kappa\mu$. Similarly, the Bingham density on Ω_q has a density

$$\frac{1}{b(K)} \exp x' K x , \qquad (3.1.2)$$

where K is a $q \times q$ symmetric matrix and

$$b(K) = \int \exp x' K x \, d\omega_q \qquad (3.1.3)$$

is seen to be a symmetric function of the eigenvalues of K . The exponent in (3.1.2) is a linear function of the elements of K . The ease of handling exponential distributions (e.g., find the maximum likelihood estimator) depends upon the complexity of the normalizing constant and we may guess that (3.1.1) is simpler to handle than (3.1.2). In Section 3.2, more general exponential families will be discussed. This will allow us to introduce the orthogonal functions on Ω_q , spherical harmonics, in terms of which we should be able to represent any density of practical interest.

An alternative to setting up a parametric family and then seeking optimal estimators and tests is setting up plausible and easily computable estimators and test statistics and seeking classes of distributions for which they will be effective and for which their properties may be derived, perhaps only in large samples. The latter is the approach of Chapters 4 and 5, and is introduced in Section 3.4.

An interesting diversion is to define an estimator and ask for what distribution(s) this estimator will be the maximum likelihood (m.l.) estimator. Thus, for example, we may ask: In what distributions on $\mathbb{R}$ with density $f(x - \mu)$ is the sample mean $\bar{x}$ always the m.l. estimator of μ ? That is, for what f is it true that

$$\prod_1^n f(x_i - \mu) \leq \prod_1^n f(x_i - \bar{x}) \ , \ \forall n \ , \quad x_1, \ldots, x_n \ . \tag{3.1.4}$$

If the maximum can be found by differentiation, this means that $\sum_1^n g(x_i - \bar{x}) = 0$ $\forall \{x_1, \ldots, x_n\}$, where $g = f'/f$. Since $\sum_1^n (x_i - \bar{x}) = 0$, we know that for any $y_1, \ldots, y_n$, $\sum_1^n y_i = 0$, $\sum_1^n g(y_i) = 0$, or that $\sum_1^{n-1} g(y_i) + g(-y_1 - \ldots -y_{n-1} =$ Take $n = 3$ so that $g(y_1) + g(y_2) + g(-y_1-y_2) = 0$ for all y_1 and y_2. Setting $y_1, y_2 = 0$, $g(0) = 0$, while $y_2 = 0$ implies that $g(y) = -g(-y)$. Hence $g(y_1) + g(y_2) = g(y_1 + y_2)$. But given only that g is measurable, a well-known result for functional equations tells us that $g(y) = f'(y)/f(y) = cy$ so that f is proportional to $\exp cy^2$. Thus if (3.1.4) is true for $n = 2$, $n = 3$, f must be the Gaussian density. An early version of this is due to Gauss. Teicher's (1961) theorem gives a slight improvement. This is just one of many *characterization theorems* for the Gaussian. A wide-ranging discussion of such theorems is given in Kagan, Linnik, and Rao (1973). It is therefore interesting to discuss such questions on Ω_q. It will be seen, for example, that in the above sense the Langevin distribution is analogous to the Gaussian.

Some statistical questions are simply answered for x in $\mathbb{R}$ if we assume that $f(x - \mu)$ is symmetric about μ, i.e., that $f(z) = f(-z)$ for all $z \in \mathbb{R}$. The Gaussian is an example. Much of the current Robustness literature is based on this assumption. Non-parametric methods may also be used then too. Then to test $\mu = \mu_0$, we could just count the number of sample members greater than μ_0 and see if it is reasonable compared to a binomial variable based on n and $p = \frac{1}{2}$. On Ω_q, an analogue would be a distribution which is rotationally symmetric about some unit vector μ. Since $x = \mu\mu' x + x_{\perp\mu}$ and the density must be the same for all $x_{\perp\mu}$, we must be able to write it as $f(\mu'x)$. If further the distribution has antipodal symmetry, then f must be an even function. Special cases are, respectively, densities proportional to $\exp k(\mu'x)$ and $\exp k(\mu'x)^2$. A more general symmetry would require that the distribution be invariant under all rotations in some s dimensional subspace V. Then, if the part of x in V is denoted by x_V, the density must have the form $f(\|x_V\|)$. These ideas are introduced in Section 3.4 for use in Chapters 4 and 5, but we will not exploit the non-parametric possibilities of these assumptions in these lectures.

By the Central Limit Theorem, the Gaussian Distribution in $\mathbb{R}$ should appear when our measurement is subject to many small independent errors of similar order of magnitude. Alternatively, if a point diffuses on $\mathbb{R}$ due to small frequent kicks coming equally likely from the left and the right, its random position at time t, after starting from μ, has a Gaussian density, mean μ and variance proportional to t. We will see in Section 3.5 that the Langevin distribution is not the analogue of the Gaussian in this sense, but that it can be obtained by other diffusions.

Section 3.6 shows two distinct classes of distributions on Ω_q. The first set are rather trivially derived from the Gaussian. The second set are derived by a much more interesting process of both practical and theoretical interest.

This chapter concludes with Section 3.7, which shows how several of the distributions may be matched with each other on the circle and the sphere, and their numerical mutual approximations.

3.2 Exponential Models and Spherical Harmonics

Suppose x has components $x_1,\ldots,x_q$. Then a natural exponential family would have density proportional to

$$\exp \sum_{j=1}^{m} G_j(x) \ , \ x\varepsilon\Omega_q \ ,$$

where $G_j(x)$ is a homogeneous polynomial in $x_1,\ldots,x_q$ of degree j. The normalizing constant would be a complicated function of the coefficients in the polynomials. In many cases, the underlying coordinate system is arbitrary in that it may be rotated so that data is represented by $y = Hx$ where H is orthogonal, without changing the underlying statistical problem. Note that the density, when expressed in terms of y, is of the same form as above, i.e., it is proportional to

$$\exp \sum_{j=1}^{m} \tilde{G}_j(y) \ , \ \tilde{G}_j(y) = G_j(H'x) \ ,$$

and $\tilde{G}_j(y)$ is also homogeneous in the components of y of degree j. Thus this form of exponential family is invariant under rotation of coordinates. It remains to determine a suitable basis for expressing homogeneous polynomials of degree j.

Following Müller, let $x\varepsilon\mathbb{R}^q$, and let $\Delta_q = \nabla\cdot\nabla = \nabla^2 = (\frac{\partial}{\partial x_1})^2 + \ldots + (\frac{\partial}{\partial x_q})^2$ be the Laplacian operator. Suppose that $G(x)$ is a polynomial, homogeneous of degree j in the components of $x\varepsilon\mathbb{R}^q$, that

$$\Delta_q G(x) = 0 \ .$$

Thus G is an harmonic function.

Let $\xi = x/\|x\|$ be a unit vector. Then, by homogeneity,

$$S_j(q,\xi) \equiv G(\xi) = \|x\|^{-j} G(x) ,$$

and S, defined on Ω_q is called a (regular) spherical harmonic of degree j in q dimensions. The term "surface harmonic" is sometimes used.

Now if $y = Hx$, H orthogonal and $f(y)$ is a function defined in $\mathbb{R}^q$, then $\frac{\partial f}{\partial x} = H \frac{\partial f}{\partial y}$ and hence the Laplacian

$$\Delta_q = \left(\frac{\partial}{\partial x}\right) \cdot \left(\frac{\partial}{\partial x}\right) = \left(H \frac{\partial}{\partial y}\right) \cdot \left(H \frac{\partial}{\partial y}\right) = \left(\frac{\partial}{\partial y}\right) \cdot \left(\frac{\partial}{\partial y}\right)$$

is invariant under rotation. Thus if $S_j(q,\xi)$ is a spherical harmonic of degree (or sometimes order) j, so is $S_j(q,H\xi)$. It can be shown that if $S_j(q,\xi)$ and $S_k(q,\xi)$ are spherical harmonics of different degrees, then they are orthogonal on Ω_q, i.e.,

$$\int_{\Omega_q} S_j(q,\xi) S_k(q,\xi) d\omega_q = 0 \qquad (j \neq k) . \tag{3.2.1}$$

It can also be shown that there are exactly

$$(2j + q - 2) \frac{(j+q-3)!}{(q-2)!j!} = N_{j,q} \tag{3.2.2}$$

linearly independent spherical harmonics of degree j, and that they form a basis for the linear space of homogeneous polynomials of degree j in the components of $x \varepsilon \Omega_q$. Moreover, they can be chosen so that they are orthogonal over Ω_q. The fact that spherical harmonics are mapped into spherical harmonics under rotation of the coordinate system shows that such a basis is mapped into another basis of the same type by rotation of the coordinates by an orthogonal matrix.

This suggests that a natural definition for an exponential family on Ω_q would be

$$(\text{function of } \beta\text{'s})\exp \sum_{j=1}^{m} \sum_{k=1}^{N_{jq}} \beta_{jk} S_{jk}(x) , \tag{3.2.3}$$

where $\{S_{jk}\}_{k=1}^{N_{jq}}$ is a complete set of orthogonal spherical harmonics of degree j on Ω_q under the rotation $y = Hx$, H orthogonal, the density can be expressed as

$$(\text{function of } \tilde{\beta}\text{'s})\exp \sum_{j=1}^{m} \sum_{k=1}^{N_{jq}} \tilde{\beta}_{jk} S_{jk}(y) ,$$

where $\tilde{\beta}_{jk}$ is a linear combination of $\beta_{j1},\ldots,\beta_{jN_{jq}}$. This density is the natural generalization to Ω_q of the exponential of a finite Fourier series on the circle.

Indeed, in two dimensions, the Laplacian expressible in polar coordinates is

$$\Delta_2 = \left(\frac{\partial}{\partial r}\right)^2 + r^{-2}\left(\frac{\partial}{\partial \theta}\right)^2 + r^{-1}\left(\frac{\partial}{\partial r}\right) ,$$

and it may be verified that

$$\Delta_2 r^j \cos j\theta = \Delta_2 r^j \sin j\theta = 0 .$$

Since $r^j \cos j\theta$ and $r^j \sin j\theta$ are expressible as homogeneous polynomials of degree j in $x_1 = r\sin\theta$ and $x_2 = r\cos\theta$, $\cos j\theta$ and $\sin j\theta$ are spherical harmonics of order j and dimension 2 . Note that $N_{j,2} = 2j \frac{(j-1)!}{j!} = 2$, and hence these form a basis. Note also that

$$\cos j\theta = T_j(\cos\theta) = T_j(\xi_2) , \tag{3.2.4}$$

where T_j is a Chebyshev polynomial of degree j which, as a function of θ , is symmetrical about $\theta = 0$, and as a function of $\xi = x/\|x\|$ is symmetric about $e_2 = [0,1]'$.

In three dimensions, a convenient basis for the spherical harmonics of order j is (expressed in spherical coordinates with $\xi_1 = \sin\theta\cos\phi$, $\xi_2 = \sin\theta\sin\phi$, $\xi_3 = \cos\theta$)

$$\begin{gathered} P_j(\cos\theta) , \cos\phi\, P_j^1(\cos\theta) , \sin\phi\, P_j^1(\cos\theta),\ldots, \\ \cos j\phi\, P_j^j(\cos\theta) , \sin j\phi\, P_j^j(\cos\theta) , \end{gathered} \tag{3.2.5}$$

where

$$P_j(x) = (2^j j!)^{-1}(\tfrac{d}{dx})^j(x^2 - 1)^j \tag{3.2.6}$$

is a Legendre polynomial of degree j , and

$$P_j^k(x) = (-1)^k(1-x^2)^{\frac{1}{2}}(\tfrac{d}{dx})^k P_j(x) , k = 1,2,\ldots,j \tag{3.2.7}$$

are associated Legendre functions. Note that there are exactly

$$N_{j,3} = (2j + 1)\frac{j!}{1!\ 1!} = 2j + 1$$

spherical harmonics of order j and dimension 3 . Note also that there is a single (and unique, up to multiplication by a scalar) spherical harmonic, namely $P_j(\cos\theta) = P_j(\xi_3)$, which is a function of $\cos\theta$ and hence is circularly symmetric about the pole $e_3 = [0, 0, 1]$.

In an arbitrary number q of dimensions, a complete representation of a basis of spherical harmonics of degree j is more complicated. Fortunately we will not need this below because we will only consider problems with certain symmetries. If e.g. $S_j(q,\xi)$ has rotational symmetry about a unit vector μ so that $S_j(q,H\xi) = S_j(q,\xi)$ for all orthogonal H leaving μ unchanged $(H\mu = \mu)$, and if $S_j(q,\mu) = 1$, Müller (op. cit.) calls $S_j(q,\xi)$ a Legendre function of degree j and writes it $P_j(q,\mu'\xi)$. Putting $t = \mu'\xi$, $P_j(q,t)$ is a polynomial in t of degree j , $P_j(q,1) = 1$, $P_j(q,t)=(-1)^jP_j(q,-t)$ As we saw above, when $q=2$ these polynomials are those of Chebyshev, and when $q=3$, they are the familiar Legendre polynomials. In general $P_j(q,t)$ satisfies

$$\left[(1-t^2)\frac{d^2}{dt^2} - (q-1)t\frac{d}{dt} + j(j+q-2)\right] P_j(q,t) = 0 \ . \qquad (3.2.8)$$

From Szegö (1939) we see that the $P_j(q,t)$ are a special case of Jacobi polynomials called ultraspherical or Gegenbauer polynomials, $C_j^{\frac{1}{2}q-1}(t)$ where

$$C_j^\lambda(t) = \frac{(-1)^j}{2^j j!}\ \frac{\Gamma(\lambda+\frac{1}{2})}{\Gamma(\lambda+\frac{1}{2}+j)}\ \frac{\Gamma(2\lambda+j)}{\Gamma(2\lambda)}\ (1-t^2)^{\frac{1}{2}-\lambda}\ \left(\frac{d}{dt}\right)^j(1-t^2)^{j+\lambda-\frac{1}{2}} . \quad (3.2.9)$$

Thus the Legendre polynomial $P_j(t) = C_j^{\frac{1}{2}}(t)$.

The Laplace operator in $\mathbb{R}^q$, expressed in polar coordinates with $r = \|x\|$, may be written as

$$\Delta_q = \frac{\partial^2}{\partial r^2} + \frac{q-1}{r}\frac{\partial}{\partial r} + \frac{1}{r^2}\Delta_q^* , \tag{3.2.10}$$

where Δ_q^* is the Beltrami operator explicitly defined in Müller (1966, p.38). Since $S_j(q,x) = r^j S_j(q,\xi)$ with $x = r\xi$, and $\Delta_q S_j(q,x) = 0$, (3.2.10) shows that

$$\Delta_q^* S_j(q,\xi) + j(j+q-2)S_j(q,\xi) = 0 . \tag{3.2.11}$$

Thus the $S_j(q,\xi)$ are the eigen functions of the Beltrami operator. If $f(\xi)$ is a continuous function on Ω_q, Müller's Theorem 9 shows that $f(\xi)$ may be approximated uniformly in the sense of

$$\lim_{r\to 1-0} \sum_{j=0}^{\infty} r^j S_j(q,\xi) = f(\xi) , \tag{3.2.12}$$

where

$$S_j(q,\xi) = \frac{N(q,j)}{\omega_q} \int_{\Omega_q} P_j(q,\xi'\eta) f(\eta) d\omega_q(\eta) . \tag{3.2.13}$$

The results of this paragraph will be used in Section 3.5.

The above discussion makes it clear that a very wide class of densities on Ω_q can be represented by Beran's (1979) class of densities.

$$c(\beta_1,\ldots,\beta_m)^{-1} \exp \sum_{j=1}^{m} \beta_j S_j(x) , \tag{3.2.14}$$

where, to simplify notation, the S_j are spherical harmonics, not necessarily of degree j, but assumed to be orthonormal so that

$$\int_{\Omega_q} S_j(x)S_k(x)d\omega_q = \delta_{jk} . \tag{3.2.15}$$

The normalization constant will ordinarily be difficult to evaluate, but the likelihood of n observations $x_1,\ldots,x_n$ may be written

$$\left.\begin{aligned} &c^{-n}\exp \sum_{j=1}^{m} \beta_j T_j , \\ &T_j = \sum_{i=1}^{n} S_j(x_i) . \end{aligned}\right\} \tag{3.2.16}$$

Hence $\{T_1,\ldots,T_m\}$ is a sufficient statistic. Then the m.l. equations for $\hat{\beta}_1,\ldots,\hat{\beta}_m$ are

$$\frac{\partial \log c}{\partial \beta_j} = \frac{T_j}{n} , \quad j=1,\ldots,m \tag{3.2.17}$$

since

$$c(\beta_1,\ldots,\beta_m) = \int_{\Omega_q} \exp(\sum_1^m \beta_j S_j(x))d\omega , \tag{3.2.18}$$

$$\frac{\partial \log c}{\partial \beta_j} = ES_j(x) , \quad j=1,\ldots,m$$

by the Law of Large Numbers,

$$T_j/n \to ES_j(x) \text{ (in prob)} , \tag{3.2.19}$$

and since c is a continuous function of its arguments

$$\hat{\beta}_j \to \beta_j \text{ (in prob) } . \tag{3.2.20}$$

(Stronger results obtain - see Barndorff-Nielsen (1978), Johansen (1979)).

Furthermore, the multivariate central limit theorem permits us to assert that the vector with elements

$$n^{\frac{1}{2}}\left(\frac{T_1}{n} - ES_1\right), \ldots, n^{\frac{1}{2}}\left(\frac{T_m}{n} - ES_m\right) \tag{3.2.21}$$

becomes asymptotically Gaussian with zero mean vector and a covariance matrix V (i.e., $G_m(0,U)$) where

$$V = [E(S_j(x)S_k(x)) - E(S_j(x))E(S_k(x))] . \tag{3.2.22}$$

By (3.2.15), if $Lx = U(\Omega_q)$, $V = I_m$. By extending the (3.2.18) argument,

$$\frac{\partial^2 \log c}{\partial\beta_j \partial\beta_k} = E\, S_j(x)S_k(x) \quad . \tag{3.2.23}$$

but

$$n^{-1} \sum_{i=1}^{n} S_j(x_i)S_k(x_i) \to E(S_j(x)S_k(x)) \text{ (in prob) },$$

so that V may be consistently estimated.

Thus the statistical theory of (3.2.14) is easy, except for the solution of (3.2.17), in large samples. An alternative method of finding efficient estimators of the β_j's is the substance of Beran's (1979) paper. The idea is to use the data to form a non-parametric estimator $\hat{f}$ of the density f. Taking logs of (3.2.14) suggests that one should then estimate $\beta_1,\ldots,\beta_m$ by treating

$$\log \hat{f}(x) = \beta + \sum_1^m \beta_j S_j(x) + \text{"error"}$$

as a linear regression problem using some form of weighted least squares. The trick is to do this, and the estimation of f, so that the resulting estimates are fully efficient.

We should note here that the $ES_j(x)$ are, in general, functions of all the β_j's, so there would be no difficulty in defining a large sample test that two populations have the same set of parameters. One simply forms a Hotelling's T^2 using (3.2.14), (3.2.16), and (3.2.18). But to test that, say, β_1 is the same in the two populations we must solve (3.2.12).

If the distribution is known to have rotational symmetries, (3.2.10) must be restricted. Antipodal symmetry, $f(x) = f(-x)$ rules out the use of $S_j(q,x)$ of odd degree, since $S_j(q,x) = (-1)^j S_j(q,-x)$. If the distribution is rotationally symmetric about a unit vector μ, then the $S_j(q,x)$ must be Legendre functions, $P_j(q,t)$. The Ph.D. Thesis of Hetherington (1981) deals with this topic, and calls upon results widely used in Theoretical Physics. As is shown in Chapters 4 and 5, it is possible to derive large sample theory for densities of the form $f(\mu' x)$, $Ex > 0$, and $Ex = 0$ for general f, and of the form $f(\|xv\|)$, so that unless a problem requires the density (3.2.10) explicitly, this theory may not be necessary.

3.3 Some Characterizations

An analogue of the Gaussian result in Section 3.1 would be: for what distributions on Ω_q with densities of the form $f(\mu'x)$ is the m.l. estimator of μ given by $\hat{\mu} = X/\|X\|$? For the circle, von Mises (1918) showed that his distribution was the only solution. Arnold (1941) proved the same result for the sphere -- that the density must be proportional to $\exp k\,\mu'x$. We now prove this result for any dimensionality q using a more powerful method.

Maximizing $\sum_1^n \log f(\mu' x_i)$ subject to $\mu'\mu = 1$, we find, setting $f'/f = g$, that $\hat{\mu}$ must satisfy

$$\sum_1^n x\, g(\hat{\mu}' x_i) = \theta\hat{\mu} \ , \tag{3.3.1}$$

where θ is a scalar Lagrangian multiplier. Since $\hat{\mu}'$ must be given by $X/\|X\|$, $X = \sum_1^n x_i$, (3.3.1) says that

$$\Sigma\, x_i \frac{g(\hat{\mu}'x)}{\|x\|} \quad \text{and} \quad \Sigma\, x_i \quad \text{are parallel,} \tag{3.3.2}$$

or setting $x_i = \hat{\mu}\hat{\mu}' x_i + x_{i\perp\hat{\mu}}$, that for all $x_i, \ldots, x_n \varepsilon \Omega_q$,

$$\Sigma\, \hat{x}_{i\perp\mu}\, g(\hat{\mu}' x_i) = 0 \ . \quad \Sigma\, x_{i\perp\hat{\mu}} = 0 \ . \tag{3.3.3}$$

Suppose $n = 3$ and that x_1, x_2, x_3 are coplanar, so we can write (3.3.3) as

$$\Sigma \sin\theta_i\, g(\cos\theta_i) = 0 \ , \quad \Sigma \sin\theta_i = 0 \tag{3.3.4}$$

or

$$\Sigma\, h(y_i) = 0 \ , \quad \Sigma\, y_i = 0 \tag{3.3.4}$$

with $y_i = \sin\theta_i$ Applying the argument in Section 3.1, we have a demonstration that $h(y_i) = k \sin\theta_i$ provided g is measurable. Hence $f'/f = g = \kappa$, so that f is proportional to $\exp \kappa \mu' x$. Hence the Langevin distribution is characterized by the requirement that the m.l. estimator of μ in $f(\mu' x)$, given samples of sizes 3 or more, be the direction of the sample resultant X, provided we assume that f'/f is measurable. Bingham and Mardia (1975) proved a stronger theorem assuming only that $f(t)$ is lower semi-continuous at $t = 1$ in complete analogy with Teicher's theorem for the Gaussian. To avoid assuming that f is differentiable means a much longer proof.

Arnold (1941) applies the idea, inherent in the above, of center of mass estimates to the semi-circle and the hemisphere to get distributions for axes or unsigned directions, and ends up with densities proportional to $\exp k|\cos\theta|$ instead of $\exp k \cos\theta$ above. The density $\exp k|\cos\theta|$, also suggested by Selby (1964), is hard to handle. However, a better analogue is the moment of inertia of the data points as mentioned in Section 1.3. This leads us to ask: for what distributions with density $f(\mu' x)$ is $\hat{\mu}_1$ (or $\hat{\mu}_q$) the m.l. estimator μ? Here we recall that $M_n = n^{-1} \sum_1^n x_i x_i'$ has eigenvectors and values $(\hat{\mu}_i, \hat{\lambda}_i)$ with $\hat{\lambda}_1 > \dots > \hat{\lambda}_q$. As before, set $f'/f = g$. Then we must have

$$\sum_1^n x_i \, g(\hat{\mu}_1' x_i) = \theta \hat{\mu}_1 , \quad M_n \hat{\mu}_1 = \hat{\lambda}_1 \hat{\mu}_1 , \tag{3.3.6}$$

where θ is a scalar Langrangian multiplier. Assume that $n = 3$, and that x_1, x_2, x_3 are coplanar. Then $\hat{\mu}_1$ (and $\hat{\mu}_2$) will lie in this plane (and $\hat{\lambda}_3 = 0$), and we may again write $\hat{\mu}_1' x_i = \cos\theta_i$, $\hat{\mu}_2' x_i = \sin\theta_i$, so that (3.3.6) may be written

$$\Sigma \sin \theta_i g(\cos \theta_i) = 0 , \quad \Sigma \sin \theta_i \cos \theta_i = 0 . \tag{3.3.7}$$

This is analogous to (3.3.4). Repeating the argument, $g(\cos \theta_i) \propto \cos \theta$. Solving $f'(z)/f(z) = g(z) = cz$, $f(z) \propto \exp \frac{c}{2} z^2$. Thus the Scheiddeger-Watson distribution is characterized by the m.l. eigenvector -- or moment of inertia-estimator property. Since eigenvectors are unsigend by definition, $f(\mu'x)$ *must* be an even function of $\mu'x$.

There are other characterizations of these densities, and also characterizations of other densities such as that of Bingham and the uniform. A survey with new results is given in Mardia (1975). We will return to this topic later but we conclude here with a result needed in the next section. M. S. Bingham (1978) and Kent, Mardia, and Rao (1979) show that among all distributions with a density, the uniform distribution is the only one for which $\|X\|$ and $X/\|X\|$ are independent. It is obvious that $\|X\|$ and $X/\|X\|$ would be independent if the distribution of x had probability $\frac{1}{k}$ on the vertices of a regular k-ogon -- that is, for a "discrete uniform" distribution.

3.4 Rotationally Symmetric Distributions

If the law of x on Ω_q is rotationally symmetric about a direction specified by the unit vector μ, we may write its density as $f(\mu'x)$ with $t = \mu'x$, set $x = \mu t + (1-t^2)^{\frac{1}{2}}\xi$, $\|\xi\| = 1$, $\xi \perp \mu$ as in (2.2.1). It follows that t and ξ are independent, ξ is uniform on the unit sphere $\{x; \|x\| = 1, x \perp \mu\}$, and the density of t is

$$\omega_{q-1} f(t)(1 - t^2)^{\frac{q-3}{2}} . \tag{3.4.1}$$

Thus

$$Ex = \mu Et + E(1-t^2)^{\frac{1}{2}} E\xi \tag{3.4.2}$$
$$= \mu Et ,$$

and

$$Exx' = E\{\mu\mu' t^2 + \mu\xi' t(1-t^2)^{\frac{1}{2}} + \xi\mu' t(1-t^2)^{\frac{1}{2}} + (1-t^2)\xi\xi'\}$$
$$= \mu\mu' Et^2 + E(1-t^2)E\xi\xi\xi' .$$

This family was just suggested by J.G. Saw (1978) and Wellner (1978). Saw suggests that pseudo samples from all distributions in this family be found by generating t's by (3.4.1) and ξ's by the methods of Section 2.6. This generalizes the method always used for the Fisher distribution. However he suggests it might be used to create correlated unit vectors, a topic we cannot discuss here.

Now in Chapter 2 we saw that if $L(x) = U(\Omega_q)$, $Exx = I_q/q$. Since ξ is uniform on the unit sphere in $\mu^{\perp}$, and the identity matrix in $\mu^{\perp}$ is $I_q - \mu\mu'$, we may assert $E\xi\xi' = (I_q - \mu\mu')/(q - 1)$, so that

$$Exx' = \mu\mu' Et^2 + E(1-t^2)(I_q - \mu\mu')/(q-1) . \tag{3.4.3}$$

As a check observe that the trace of both sides of (3.4.3) is unity.

Given n independent copies $x_1, \ldots, x_n$ of x, $X = x_1 + \ldots + x_n$, the density of $\mu'X$ is obtained by Fourier inversion of the nth power of

$$\psi_f(\theta) = \omega_{q-1} \int_{-1}^{1} e^{i\theta t} f(t)(1-t^2)^{q-3/2} dt \tag{3.4.4}$$

In Chapter 2, we studied this when $f(t) = \omega_q^{-1}$, and used the facts that then $X / \|X\|$ and $\|X\|$ are independent and $X/\|X\|$ uniform on Ω_q to get the distribution of X. Those results no longer apply in general, as we saw at the end of the last section. If f is proportional to $\exp k\,\mu'x$, it is possible to finite sample results. However it is possible to get asymptotic results from any density $f(\mu'x)$.

By the C.L.T.,

$$\mathcal{L} n^{\frac{1}{2}} (n^{-1}X - \mu Et) \to G_q(0,V) \tag{3.4.4}$$

where

$$\begin{aligned} V &= Exx' - (Ex)(Ex)' \\ &= \mu\mu' \operatorname{var} t + E(1-t^2)(I_q - \mu\mu)/(q - 1). \end{aligned} \tag{3.4.5}$$

Furthermore, the W.L.L.N. implies

$$n^{-1} \sum_1^n x_i x_i' - XX'/n^2 \to V \text{ (in prob) as } n \to \infty . \tag{3.4.6}$$

Thus the behaviour of X in large samples is the same for all rotationally symmetric distributions having the same first and second moments, Et, Et^2, where $t = \mu'x$.

If $f(t)$ has a maximum at $t = 1$ and falls off steadily as t decreases to -1, the mode of $f(\mu'x)$ is at $x = \mu$. This means $f'(t) \geqslant 0$.

$$E(t) = \omega_{q-1} \int_{-1}^{1} tf(t)(1 - t^2)^{q-3/2} dt$$

$$= \omega_{q-1}/q-1 \int_{-1}^{1} f'(t)(1 - t^2)^{q-1/2} dt > 0$$

unless $f'(t) = 0$. Then $EX = \mu n E(t)$ points in the direction μ. Whenever $Et > 0$, this will be so. Hence $X/\|X\|$ has some merit then as an estimator of μ, and we will be able to base statistical methods on X and use (3.4.4). This topic will be pursued later in Chapter 4.

If $Et = 0$, X will not be a useful estimator of μ. Then the fact, seen from (3.4.5), that V has an eigenvector μ, eigenvalue var t, and $(q - 1)$ equal eigenvalues $(1 - Et^2)/(q - 1)$ associated with the eigen subspace $\mu^{\perp}$, suggests that we should estimate μ by computing the eigenvalues of M_n. They will be distinct with probability one (see, e.g., Okamoto (1973)), but if n is large and V is true, there will be an isolated root and a cluster of $q-1$ similar roots provided var$t \neq (1-Et^2)(q-1)^{-1}$ The eigenvector of the isolated root should be a good estimator of μ. This topic will be pursued later, in general and for the special case of f proportional to $\exp k\,(x'\mu)^2$ in Chapter 5.

Rotational symmetry about a unit vector μ may be generalized. Let $\mathcal{V}$ be a subspace of $\mathbb{R}^q$ of dimension s. Then $x \varepsilon \Omega_q$ may be written

$$x = x_v + x_{\perp v} ,$$

where $x_v \,\varepsilon \mathcal{V}$ and $x_{\perp v} \,\varepsilon \mathcal{V}^{\perp}$. Then if $f(x)$ depends only upon x_v, not upon $x_{\perp v}$, the distribution is rotationally symmetric within V. A subclass of such distributions would have $f(x)$ dependent only upon

$\|x_v\|$. This would have antipodal symmetry too since $\|x_v\| = \|-x_v\|$. A special case of interest is the density proportional to

$$\exp k \|x_v\|^2 . \tag{3.4.7}$$

This could be called the gS-W (generalized Scheidegger-Watson) distribution. By the previous arguments, if the x is rotationally symmetric within $\mathcal{V}$, $x \varepsilon \Omega_q$,

$$\left.\begin{aligned} &Ex \; \varepsilon \mathcal{V} \\ &Exx' = \frac{c}{s} P_v + \frac{1-c}{q-s} (I_q - P_v) , \end{aligned}\right] \tag{3.4.8}$$

where $P_v = P_v' = p_v^2$ is the orthogonal projector onto $\mathcal{V}$ and $1>c>0$. For future reference, we note the generalization of (2.2.1) and (2.2.2)

$$\left.\begin{aligned} x &= \|x_v\| \xi + (1-\|x_v\|^2)^{\frac{1}{2}} \eta , \quad \xi \varepsilon \mathcal{V}, \quad \eta \varepsilon \mathcal{V}^{\perp} \\ d\omega_q &= t^{s-1}(1 - t^2)^{\frac{q-s}{2} -1} \quad dt \, d\omega_s \, d\omega_{q-s} , \; t = \|x_v\| . \end{aligned}\right\} \tag{3.4.9}$$

Since $t = \|x_v\|$, $0 \le t \le 1$, $\xi(\eta)$ ranges over an Ω_s (Ω_{q-s}) .
The statistical problems with such distributions will begin with the estimation of $\mathcal{V}$. From (3.4.8), if $Ex = 0$, we will naturally look at $M_n = n^{-1} \Sigma x_i x_i'$ and seek its eigenvectors associated with a cluster of s , similar eigenvalues. Even if $Ex \neq 0$, $n^{-1}X$ will not be enough to estimate $\mathcal{V}$ when $s > 1$. We must estimate P_v . We will return to this topic in Chapter 5.

As stated in the introduction, the non-parametric possibilities of these assumptions will not be exploited. But they are obvious. For example, if we have samples from two populations with densities, rotational symmetry about μ and with $Ex>0$, and wish to test that they are identical, except possibly have different μ , the permutation distribution of $\|x_1\| + \|x_2\| - \|x_1 + x_2\|$ could be used. With a computer, this is a practical procedure. The theory might be tedious.

It is seen above that the assumption of rotational symmetry gives us a natural quantity to estimate -- the direction μ or, more generally, the subspace $\mathcal{V}$. Further, the expectation and dispersion matrix of the random variable x take simple forms. Given the C.L.T. for vectors $x \in \Omega_q$, large sample theory should then be easy to develpp -- this is the substance of Chapters 4 and 5. But given large samples, a more general theory for the case $Ex \neq 0$ could be based on

$$\mathcal{L} n^{\frac{1}{2}}(n^{-1}X - Ex) \to G_q(0,\Sigma) , \tag{3.4.10}$$

where Σ is the dispersion or variance-covariance matrix of x , which is always here consistently estimated by

$$S = n^{-1} \sum_1^n x_i x_i' - n^{-2} XX'.$$

If we write

$$\left.\begin{aligned} &Ex = c\mu \\ &0 < c < 1 , \|\mu\| = 1 , \end{aligned}\right\} \tag{3.4.11}$$

then c is a measure of the concentration of the distribution on Ω_q about the "preferred" direction μ . The basic result is that

$$\mathcal{L} n (n^{-1}X - c\mu)' S^{-1}(n^{-1}X - c\mu) \to \chi_q^2 , n \to \infty . \tag{3.4.12}$$

Since $\|X\|n^{-1}$ is a consistent estimator of c, we may use it in the l.h.s of (3.4.12) to derive a confidence region for the preferred direction μ, or a test of a specified μ.

We may rewrite (3.4.9) as

$$n^{\frac{1}{2}}\left(\frac{X}{\|X\|} - \mu\right) \sim G_q(0, \frac{\Sigma}{c^2}) , \tag{3.4.13}$$

Thus, given a sample of n_i with resultant X_i $(i = 1,2)$, to test $\mu_1 = \mu_2$, we would naturally look at

$$\hat{\mu}_1 - \hat{\mu}_2 = \frac{X_1}{\|X_1\|} - \frac{X_2}{\|X_2\|} \tag{3.4.14}$$

which, if $\mu_1 = \mu_2$, is asymptotically distributed as

$$G_q\left(0, \frac{\Sigma_1}{n_1 c_1^2} + \frac{\Sigma_2}{n_2 c_2^2}\right). \tag{3.4.15}$$

The covariance matrix in (3.4.15) could be approximated by

$$S = \frac{S_1}{n_1\|X_1/n_1\|^2} + \frac{S_2}{n_2\|X_2/n_2\|^2} , \tag{3.4.16}$$

so that

$$\mathcal{L}(\hat{\mu}_1 - \hat{\mu}_2)^1 \, S^{-1}(\hat{\mu}_1 - \hat{\mu}_2) \sim \chi_q^2 \tag{3.4.17}$$

as $n \to \infty$.

We will not further develop this line of argument. Nor will we consider general procedures for the case $Ex = 0$.

3.5 Distributions Arising From Diffusion

We saw in Sections 3.2 and 3.3 that the Langevin is, in several senses, an analogue of the Gaussian. The Gaussian also arises by addition and scaling, as envisaged in the Central Limit Theorem. On the circle one could use addition of angles modulo 2π. Because there seems to be no natural scaling, this gives the uniform distribution as the limit, except in certain degenerate cases. It is less obvious on the sphere what to do. However, Brownian motion is much more successful.

If a particle moves on $\mathbb{R}^1$ with independent steps $dx = \xi dt$, $E\xi = 0$, $E\xi^2 = \sigma^2$ in time steps dt, the probablilty density $\phi(x,t)$ of its position at time t, when it starts at $x = 0$, is clearly

$$\phi(x,t) = (2\pi)^{-\frac{1}{2}}(\sigma^2 t)^{-\frac{1}{2}}\exp(-x^2/2\sigma^2 t) \quad . \tag{3.5.1}$$

Suppose now that the same motion occurs on the circumference of a circle of unit radius. Then the density of the position θ, $0 \leq \theta \leq 2\pi$, of the particle at time t is given by

$$f(\theta,t) = \sum_{k=-\infty}^{\infty} \phi(\theta + 2k\pi,t) \; , \tag{3.5.2}$$

because the linear positions $\theta + 2k\pi$, are all equivalent on the circle. The r.h.s. of (3.5.2) has period 2π, and a simple calculation using the characteristic function of the Gaussian shows that the Fourier representation of $f(\theta,t)$ is

$$f(\theta,t) = \frac{1}{2\pi}\sum_{m=-\infty}^{\infty} \exp\left(\frac{-m^2\sigma^2 t}{2}\right) \exp im\theta \; , \tag{3.5.3}$$

$$= \frac{1}{2\pi}\left(1 + 2\sum_{m=1}^{\infty} \exp\left(\frac{-m^2\sigma^2 t}{2}\right) \cos m\theta\right) \tag{3.5.4}$$

Because of (3.5.2), $f(\theta,t)$ is often called the "rolled-up" or "wrapped" normal density.

The von Mises density with mode at $\theta = 0$ and concentration κ has the Fourier expansion

$$\frac{\exp(\kappa\cos\theta)}{2\pi I_o(\kappa)} = \frac{1}{2\pi}\ (1 + 2 \sum_{m=1}^{\infty} \frac{I_m(\kappa)}{I_o(\kappa)} \cos m\theta)\ , \tag{3.5.5}$$

where $I_m(\kappa)$ is a modified Bessel function -- see e.g., Abramowitz and Stegun (1965, §9.6). This is clearly not the same as (3.5.4), and yet the numerical agreement is remarkable if we match by setting

$$e^{-\sigma^2 t/2} = I_1(\kappa)/I_o(\kappa) \tag{3.5.6}$$

-- see Figures 3.1, 3.2, 3.3.

The original calculations were made by Stephens (1963), prompted by the work of Ursell and Roberts (1960) on the relationship of Fisher's distribution on the sphere and the Brownian distribution on the sphere. To find the generalization of (3.5.4) on Ω_q, we could follow Ursell and Roberts (op. cit.), or argue roughly as follows. If x moves in $\mathbb{R}^q$ in steps $dx = \xi dt$, where $\xi \in \mathbb{R}^q$ has $E\xi = 0$, $E\xi\xi' = I_q\sigma^2 dt$, and the density of position at time t is $f(x,t)$, then

$$\begin{aligned} f(x,t + dt) &= E\ f(x + \xi dt) \\ &= f(x,t) + dt\ E(\xi'\Delta f) \\ &\qquad + \frac{dt}{2}\ E(\xi'\Delta\Delta'\xi) + o(dt)\ , \end{aligned}$$

so that as $dt \to 0$,

$$\frac{\partial f}{\partial t} = \frac{\sigma^2}{2} \Delta^2 f . \tag{3.5.7}$$

If this motion is restricted to the surface Ω_q , we would expect (3.5.7) would become

$$\frac{\partial f}{\partial t} = \frac{\sigma^2}{2} \Delta_q^* f , \tag{3.5.8}$$

where Δ_q^* is the Beltrami operator defined by (3.2.6) and in Müller (op. cit.). Supposing that the particle starts at $\mu \varepsilon \Omega_q$, then the symmetry of the motion implies that the distribution of position should be rotationally symmetrec about μ . To solve (3.5.8) by separation of variables, set $f = T(t)\ X(x)$, $x \varepsilon \Omega_q$ and observe that we must have

$$\frac{2}{\sigma 2} \frac{\partial T}{\partial t} \frac{1}{T} = \lambda = \frac{1}{X(n)} \Delta_q^* X(x) . \tag{3.5.9}$$

From (3.2.7), we must set

$$\lambda = -j(j + q - 2) , \quad X(x) = S_j(q,x) \tag{3.5.10}$$

and

$$T = \exp - \frac{\sigma^2}{2} j(j + q - 2)t . \tag{3.5.11}$$

But the rotational symmetry implies that we must choose

$$S_j(q,x) = P_j(q,\mu' x) . \tag{3.5.12}$$

To get a solution,we may superpose such solutions.

The particle must initially be at μ, so we need to find coefficients c_j so that

$$\sum_{j=0}^{\infty} c_j \exp(-\frac{\sigma^2}{2} j(j + q - 2)t) P_j(q,\mu'x)$$

tends to the delta function as $t \to 0$. It is then found that

$$f(x_1t) = \frac{1}{\omega_q} \sum_{j=0}^{\infty} N(q,j) \exp(-\frac{\sigma^2}{2} j(j + q - 2)t) P_j(q,\mu'x) . \qquad (3.5.13)$$

This reduces when $q = 2$ tp (3.5.4).

Ursell and Roberts (op. cit.) showed that (3.5.13) could, for $q = 3$, be well matched with the Fisher distribution. For further calculations, see Figures 3.4, 3.5, 3.6.

It was shown by Hartman and Watson (1974) that it is possible to stop this diffusion at a random time so that the distribution of positions of the particle when stopped is precisely the Langevin distribution on Ω_q, which has the density

$$\frac{\kappa^{(q/2)-1}\exp(\kappa\mu'x)}{(2\pi)^{q/2} I_{(q/2)-1}(\kappa)} . \qquad (3.5.14)$$

Pitman and Yor (1980) have shown that there is more than one such stopping time distribution.

It is also possible to get densities proportional to $\exp \kappa\mu'x$ and $\exp \kappa(\mu'x)^2$ precisely as limiting distributions of other diffusion models. These limiting distributions can sometimes be found very simply by using the Boltzmann distribution, described in books on Statistical Mechanics. If $f(x_1, \ldots, x_m)$ is the probability density of particles in a phase space $\mathbb{R}^m$, $E(x_1, \ldots, x_m)$ the energy of a particle with these coordinates in the phase space, and the average energy

$$\int E(x)p(x)dx = \text{constant},$$

then the density that maximizes the entropy is

$$f \text{ proportional to } \exp(\lambda E) .$$

In physical situations $\lambda = 1/kT$, where k is Boltzmann's constant and T is the absolute temperature. Langevin (1905) used this result to derive the angular distribution of magnetic dipoles, strength m, in a parallel magnetic field H when jostled by molecular vibrations and/or Brownian motion. The work done (and so energy acquired) to turn a dipole an angle θ away from the field is $mH\cos\theta$, so he derived Fisher's distribution with $\kappa = mH/kT$. By inventing more complicated energies, we can get any of the exponential family this way, e.g., $\exp \kappa(\mu'x)^2$.

This is a <u>macro-scopic</u> argument. There is a parallel micro-scopic argument -- follow the motion of one particle. It leads to a stochastic differential equation that is also named after Langevin. Paralleling our derivation of (3.5.1), imagine a diffusion on a circle where the step $d\theta$ has a drift term $\sin^2\theta\, dt$. Then

$$d\theta = \sin^2\theta dt + \xi\sqrt{dt} ,$$

a Langevin equation. The drift will tend to concentrate the particle near $\theta = 0,\pi$, so it is not surprising that the limiting distribution is proportional to $\exp \kappa\cos^2\theta$. Kent (1976) used the same argument to get $\exp \kappa\cos\theta$.

Another macroscopic picture that leads to these results on the circle is the diffusion of particles in a circular pipe of cross section A. Let the equilibrium concentration of particles be $f(\theta)$. Then by Fick's Law, the diffusion across A at θ is $-AD\partial f/\partial\theta$, where D is a diffusion coefficient. If the medium has velocity $v(\theta)$ at θ , the transport across A is $Av(\theta)f(\theta)$. At equilibrium, the number of particles between θ and $\theta + d\theta$ must be constant, so that

$$-AD \left.\frac{\partial f}{\partial\theta}\right|_{\theta} + AD \left.\frac{\partial f}{\partial\theta}\right|_{\theta+d\theta} + Afv\Big|_{\theta} - Afv\Big|_{\theta+d\theta} = 0 ,$$

or, for constant A,D ,

$$D \frac{\partial^2 f}{\partial\theta^2} = \frac{\partial}{\partial\theta}(fv) . \tag{3.5.15}$$

If the total flow around the tube is zero, and $v = -\kappa\sin\theta$, then (3.5.15) has the solution f proportional to $\exp \kappa\cos\theta$. A similar physical picture could be constructed by using the conduction and transport of heat. To get different laws one only needs to vary $v(\theta)$.

In the discussion of Kendall (1974), Reuter suggested a different kind of diffusion process. Let (Brownian) particles be steadily released from the origin of the sphere $\|x\| \leqslant 1$, and record where they first hit Ω_q . The distribution of first hits will be uniform if there is no drift by symmetry, but if there is a constant drift within Ω_q parallel to μ , the hit density is proportional to $\exp \kappa\mu'x$.

First we give a macroscopic proof. Let f be the equilibrium concentration of particles at any point in or on the sphere when they are steadily produced at the rate of 1 per unit time at the origin. Let them diffuse (but not interact with each other) in a medium that moves with an arbitrary velocity $v(x)$. The answer we seek is $\partial f/\partial n|_{\Omega_q}$, the normal derivative of f on $x = 1$.

If V is any small region with volume $|V|$ within the sphere with boundary ∂V, the loss of particles from V due to transport is

$$\int_{\partial V} n\cdot(vf)dS \cong |V|\operatorname{div}(vf) , \quad |V| \to 0 ,$$

where n is an outward normal and dS an area element on ∂V. The gain due to diffusion is

$$\int_{\partial V} \frac{\partial f}{\partial n} dS \cong |V|\nabla^2 f \qquad |V| \to 0 .$$

If V is a vanishingly small region that does not include the origin and equilibrium is attained, the concentration f satisfies

$$\begin{aligned} \nabla^2 f - \operatorname{div} v\, f &= 0 , && \|x\| \neq 0,1 \\ f &= 0 , && \|x\| = 1 . \end{aligned} \tag{3.5.16}$$

This is, of course, a generalization of (3.5.15). To ensure a suitable source of particles at the origin we must demand that

$$\begin{aligned} f &\to \frac{1}{\omega_q(q-2)} \frac{}{r^{q-2}} && (q \neq 2) , \\ &\to \frac{1}{2\pi} \log \frac{1}{r} && (q = 2) , \end{aligned} \tag{3.5.17}$$

as $r = \|x\| \to 0$.

For the special case $q = 2$, $v_x = c$, $v_y = 0$, we may try $f = \exp(kx)g(x,y)$ in (3.5.16). It is readily seen that if we choose $k = c/2$, g must satisfy

$$\nabla^2 g = \frac{c^2}{4} g ,$$

$$g = 0 , \quad r^2 = x^2 + y^2 = 1 , \tag{3.5.18}$$

$$g \sim \frac{1}{2} \log \frac{1}{r} , \quad r \to 0 .$$

But (3.5.18) implies that g is a function only of r. Hence

$$f(x,y) = g(r) \exp \frac{c}{2} r\cos ,$$

so that $\partial f/\partial r$ on the boundary $r = 1$ is proportional to $\exp(c/2 \cos\theta)$. Inspection shows that the proof trivially extends to any number q of dimensions. Finally, by choosing other velocity fields, other distributions on the sphere may be obtained. This is technically difficult unless the velocity fields are generated by a potential as in classical hydrodynamics.

A proper probabilistic proof uses the Cameron-Martin theorem -- see, e.g., McKean (1969, p. 97). A proof along these lines for $q = 2$ was given by Gordon and Hudson (1977); a heuristic version of this proof may be of interest. Suppose a Brownian particle is released from the origin in a plane and subject to drift along the x-axis. At each time interval dt, it moves a step $(dx + dt, dy)$ where all dx and dy are independent and Gaussian with means zero and variances dt. Thus the x-coordinate at time t is $X\ (t) = X(t) + \delta t$, which is Gaussian. Let T be the first time the particle hits the circle $x^2 + y^2 = 1$. Then $X\ (T) = \cos\theta$, where θ is the angle between the unit vectors $(X\ (T)), Y(T))$ and $(0,1)$. We would know the distribution of this angle if we knew $Eg(\cos\theta)$ for all functions g.

If $\delta = 0$, symmetry shows that θ will be uniformly distributed on $(0, 2\pi)$. Further, T and θ should be independent.

Parenthetically, we note that if Z is a random variable with a density $f(z,\alpha)$, then

$$E(Z|\alpha) = \int z\, f(z,\alpha)dz ,$$

$$= \int z \frac{f(z,\alpha)}{f(z,\alpha_0)} f(z,\alpha_0)dz , \tag{3.5.19}$$

$$E(Z|\alpha) = E(Z \frac{f(Z,\alpha)}{f(Z,\alpha_0)} |\alpha = \alpha_0$$

The ratio in (3.5.19) is called a likelihood ratio.

In our problem we know how to find the expectation of $g(\cos\theta)$ when $\delta = 0$, which we would like to correspond to θ_0 in (3.5.19). What corresponds to the likelihood ratio? Our random variable is more complicated -- it is a random process. When $\delta = 0$, the process may be thought of as a set of n steps dx, independent and $G_1(0,dt)$, so their likelihood ought to be

$$\Pi \frac{1}{(2\pi dt)^{n/2}} \exp -\tfrac{1}{2}\Sigma \frac{(dx-0)^2}{dt} , \tag{3.5.20}$$

while the likelihood of the steps dX_q, independent $G_1(\delta dt,dt)$, ought to be

$$\frac{1}{(2\pi dt)^{n/2}} \exp -\tfrac{1}{2}\Sigma \frac{(dx-\delta dt)^2}{dt} . \tag{3.5.21}$$

The ratio of (3.5.21) to (3.5.20) is $(\Sigma dx = x(t)$, $\Sigma dt = ndt=t$,

$$\exp\ (\delta X(t) - \frac{\delta^2}{2} t)\ . \tag{3.5.22}$$

The expectation of the r.h.s. is to be taken remembering that $X(t)$ is the undrifted process where θ is uniformly distributed and where our intuition would suggest that θ and T, the hitting time, are independent. Hence we may write for any function g,

$$Eg(X_\delta(T)) = \int_0^{2\pi} g(\cos\theta)\exp\delta\cos\theta\ \frac{d\theta}{2\pi}\ E(\exp\frac{-\delta^2 T}{2})\ , \tag{3.5.23}$$

so that the density of θ is proportional to $\exp\delta\cos\theta$ -- that is, has the von Mises distribution. The idea of combining (3.5.19) and (3.5.22) is the essence of the Cameron-Martin theorem. See Section 3.6 for another proof.

In our analysis above, we have used particle diffusion, but heat conduction is the more usual analogy and was used by Arnold (1941). He imagined one unit of heat placed on the spherical (or circular) surface, and computed the temperature distribution at time t later. Temperature and probability satisfy the same diffusion equation. Arnold obtained (3.5.4) and (3.5.13) with $q = 3$; to get axial distributions he released equal quantities of heat at the opposite ends of a diameter. For (3.5.4), this leads to the density $\frac{1}{2}[(f(\theta,t) + f(\theta + \pi t)]$ or

$$\frac{1}{2\pi}(1+2\sum_1^\infty \exp(-2m^2\delta^2 t)\ \cos 2m\theta)\ , \tag{3.5.24}$$

and a similar result for (3.5.13). Clearly one could get other densities by releasing unequal quantities of heat at antipodal -- or even arbitrary -- points.

3.6 Other Methods

Let y be a random vector in $\mathbb{R}^q$ with density $f(y)$, and set $y=r\ell$, where $r = \|y\|$, $\|\ell\| = 1$ so that $\ell\varepsilon\Omega_q$. Then the density of ℓ is

$$g(\ell) = \int_0^\infty f(r\ell)r^{q-1}dr \ , \qquad (3.6.1)$$

the marginal distribution of the direction of a vector.

For example if $L(y) = G_q(\mu,\Sigma)$, the density g is here called the angular Gaussian and special cases have special names. If $\Sigma = \sigma^2 I_q$, $\lambda = \mu / \|\mu\|$, $\cos\theta = \ell'\lambda$, the joint density of r and ℓ is

$$\frac{r^{q-1}}{\sigma^q(2\pi)^{q-2}} \exp\{-\frac{1}{2\sigma^q}(r^2 + \|\mu\|^2)\} \exp\{\frac{r\|\mu\|}{\sigma^2}\lambda'\ell\} \ . \qquad (3.6.2)$$

Setting $s = r/\sigma$, $m = \|\mu\|\sigma$, the density of ℓ is

$$g_q(\ell;m,\lambda) = \frac{1}{(2\pi)^{q/2}} \int_0^\infty \exp\{-\frac{1}{2}(s^2+m^2-2sm\lambda'\ell)\}s^{q-1}ds \ . \qquad (3.6.3)$$

Series expansions of (3.6.3) for $q=2$ and $q=3$ are given in the next section. (For general results, see Appendix C.) These densities for $q=2$ and $q=3$ are compared numerically with the von Mises and Fisher densities in Watson (1982). Again it is possible for any κ to find, by choice of m, an angular Gaussian distribution which matches it very well - see, e.g., Figs. 3.7, 3.8, 3.9.

Fisher (1953) observed that his distribution could be obtained by <u>conditioning</u> $L(y) = G_q(\mu,\sigma^2 I)$. For the conditional distribution of y, given $\|y\| = 1$, is clearly from (3.7.2) proportional to $\exp\kappa\lambda'\ell$ where $\kappa = r\,\|\mu\|\,\sigma^2$. It is also obvious how one could get the Bingham distribution by conditioning $G_q(0,\Sigma)$ to $\|y\| = 1$, and the Fisher-Bingham distribution (Kent (1982)) by using $G_q(\mu,\Sigma)$.

Consider again the diffusion leading to (3.5.23). The coordinates of the position $(X(t),Y(t))$ of such a particle are independent, $\mathcal{L}X(t)=G_1(\delta t,t)$, $\mathcal{L}Y(t)=G_1(0,t)$. Given $X^2(t)+Y^2(t)=1$, the density on this unit circle is proportional to $\exp\delta\cos\theta$. Since t does not appear, this result holds for $t=T$, the first hitting time, giving another proof of (3.5.23).

We turn now to an entirely different line of thought. The following process might however often lie behind directional data. Geologists often try to determine how a rock has been deformed by studying the deformation of embedded objects. The simplest deformation is called "homogeneous strain" and is simply a linear transformation, i.e., a point z in $\mathbb{R}^q$ is moved to Bz where B is a $q \times q$ matrix, $\det B > 0$. Watson (1982) describes this process and gives the following generalization of an argument due to Owens (1973).

If $g(\ell)$ denotes the density of the direction of line like objects, the density of their new directions $m = B\ell/\|B\ell\|$ is given by

$$h(m) = \frac{1}{\det B}\,\frac{g(B^{-1}m/\|B^{-1}m\|)}{\|B^{-1}m\|^{q}} \quad . \tag{3.6.4}$$

The right-hand side of this is unchanged if $B \to kB$, so we may assume $\det B = 1$. (This means that one cannot determine the dilatation $\det B$ of the strain from data on lines.)

A most elegant formula arises if g is ω_q^{-1} (i.e., the lines are initially uniform):

$$\det B = \omega_q^{-1} \int_{\Omega_q} \|B^{-1}m\|^{-q} d\omega_q \quad . \tag{3.6.5}$$

There are many more fascinating formulae of this type.

We can rewrite (3.6.4) when $g = \frac{1}{\omega_q}$ and $B^{-1} = T$ as

$$\frac{\det T}{\omega_q} \|T\ell\|^{-q} = \frac{\det T}{\omega_q} \frac{1}{(\ell' T'T\ell)^{q/2}} \quad . \tag{3.6.6}$$

If we set $T'T = \Sigma^{-1}$, Σ positive definite, (3.6.6) becomes

$$\left[\omega_q (\det\Sigma)^{\frac{1}{2}} (\ell'\Sigma^{-1}\ell)^{q/2} \right]^{-1} , \tag{3.6.7}$$

which has been noted by several authors (e.g., King (1980)). It is the marginal distribution of ℓ if $r\ell$ is $G_q(0,\Sigma)$, and so a special case of the angular Gaussian distribution.

It is obvious that the class of distributions (3.6.4) is closed under linear transformations - as is the class of Gaussian distributions (among others) in $\mathbb{R}^q$ - unlike the angular Gaussian, Langevin or Brownian distributions mentioned earlier in this chapter. We will describe the statistical theory of (3.6.6) and its applications elsewhere. It must be obvious now, that it is rather silly to speak of <u>the</u> analogue of the Gaussian in other spaces.

3.7 Arithmetic and Graphical Comparisons of Distributions

The definitions of the von Mises-Fisher, Brownian, and angular Gaussian densities only suggest that they may be numerically similar when their concentration parameters are either very small or very large. If they can always be matched very closely by appropriate choices of their concentration parameters, then for statistical analysis it should matter little which is used even if a specific family is indicated by stochastic modelling. We may then use the distribution which is most convenient. For example, it is much easier to simulate the Brownian and angular Gaussian on the circle than the von Mises distribution since they require only the generation of one and two Gaussians, respectively, and simple algorithms. On the sphere, the angular Gaussian and Fisher distributions are easier than the Brownian since they require only three Gaussians and three uniforms, respectively, and simple algorithms. For statistical mathematics, other "rankings" obtain. Of course the distributions of some statistics calculated from samples from different parents might be rather different even if the parent populations are "well" matched, i.e., could not be distinguished with available data. Such statistics should not be used since they are too sensitive to parental form. Hence the arithmetical comparisons, of this section have many uses.

To facilitate the comparison we gather up previous formulae. The angular Gaussian densities in 2 and 3 dimensions are

$$ag_2(\ell,\lambda,m) = \frac{1}{2\pi}\int_0^\infty \exp\{-\tfrac{1}{2}(s^2 + m^2 - 2sm\ell'\lambda)\}s\,ds \tag{3.7.1}$$

$$= \frac{1}{2\pi}\{g_0(m) + 2\sum_1^\infty g_p^{(2)}(m)\cos p\theta\}\ , \tag{3.7.2}$$

where $\cos\theta = \ell'\lambda$ and

$$g_p^{(2)}(m) = \int_0^\infty I_p(sm) \exp\{-\tfrac{1}{2}(s^2 + m^2)\}s\,ds , \tag{3.7.3}$$

on using the series expansion (3.7.8). The same expansion could be used to express

$$ag_3(\ell,\lambda,m) = \frac{1}{(2\pi)^{3/2}} \int_0^\infty s^2 \exp\{-\tfrac{1}{2}(s^2 + m^2 - 2sm\ell'\lambda)\}ds \tag{3.7.4}$$

as a series in $\cos p\theta$. We however need an expansion in $P_p(\cos\theta)$,

$$ag_3(\ell,\lambda,m) = \sum_0^\infty \frac{2p+1}{4\pi} g_p^{(3)}(m) P_p(\cos\theta) . \tag{3.7.5}$$

It is known that (see e.g. Abramowitz and Stegun (1965, 10.2-36)) that

$$e^{z\cos\theta} = \sum_{p=0}^\infty (2p+1)\left(\frac{\pi}{2z}\right)^{\frac{1}{2}} I_{p+\frac{1}{2}}(z) P_p(\cos\theta) . \tag{3.7.6}$$

Putting this expansion in (3.7.4), we find

$$g_p^{(3)}(m) = \int_0^\infty s^{3/2} m^{-\frac{1}{2}} I_{p+\frac{1}{2}}(sm) \exp\{-(s^2 + m^2)/2\}ds . \tag{3.7.7}$$

The von Mises and Fisher densities have, respectively, the expansions

$$\begin{aligned} vm(\theta,\kappa) &= \frac{1}{2\pi I_0(\kappa)} \exp \kappa \cos\theta \\ &= \frac{1}{2\pi}\left(1 + 2\sum_1^\infty \frac{I_p(\kappa)}{I_0(\kappa)} \cos p\theta\right) , \end{aligned} \tag{3.7.8}$$

and with $\ell'\lambda = \cos\theta$ and using again (3.7.6),

$$f(\ell,\lambda,\kappa) = \frac{\kappa}{4\pi\sinh\kappa} \exp\kappa\ell'\lambda$$

$$= \sum_{p=0}^{\infty} \frac{2p+1}{4\pi} \left(\frac{\pi\kappa}{2}\right)^{\frac{1}{2}} \frac{I_{p+\frac{1}{2}}(\kappa)}{\sinh\kappa} P_p(\cos\theta) \; . \quad (3.7.9)$$

The last formula has a better appearance if $\sinh\kappa$ is expressed in terms of $I_{\frac{1}{2}}(\kappa)$. The Brownian densities in 2 and 3 dimensions are

$$br_2(\theta, v) = \sum_{1}^{\infty} \frac{1}{\sqrt{2\pi v}} \exp - \frac{1}{2}\frac{(\theta-2\pi p)^2}{v} \quad (3.7.10)$$

$$= \frac{1}{2\pi}(1+2 \sum_{1}^{\infty} \exp(\frac{-p^2 v}{2}) \cos p\theta) \quad (3.7.11)$$

and

$$br_3(\theta, \phi, v) = \frac{1}{4\pi} \sum_{p=0}^{\infty} (2p+1)\exp(-p(p+1)v/4)\, P_p(\cos\theta) \; . \quad (3.7.12)$$

Kendall (1974) plotted sheaves of von Mises and Brownian densities and one $E\cos\theta$ matched pair of densities. In the discussion of his paper, John Kent showed graphs of $E\cos\theta$ matched triplets -- von Mises, Lack (derived from a bird model), and Brownian densities. The peaks at $\theta=o$ and in the tails are in that order, while in the middle of the range it reverses. The degree of agreement depends upon κ since the agreement becomes perfect as $\kappa \to o$ and $\kappa \to \infty$.

We wish to add the angular Gaussian to those comparisons, and to consider the spherical versions as well. Further $E\cos\theta$ matching may be supplemented by other methods, e.g.,

(a) eye matching of densities (subjective);

(b) matching so that the supremum of the difference of the cumulative distributions of θ was a minimum (objective, obtained a computer search).

We begin with the circular versions. As mentioned earlier, Stephens made the Brownian - von Mises $E\cos\theta$ comparisons.

To match the circular angular Gaussian to the von Mises by equating values of $E\cos\theta$, m and κ must be related by

$$\frac{I_1(\kappa)}{I_0(\kappa)} = \int_0^\infty \exp\left\{-\frac{1}{2}(s^2+m^2)\right\} I_1(sm)s\,ds. \tag{3.7.13}$$

From asymptotic expansions, we have

$$\begin{aligned}\frac{I_1(\kappa)}{I_0(\kappa)} &\sim \frac{\kappa}{2}\left(1 - \frac{\kappa^2}{8}\right), \quad \kappa \to 0, \\ &\sim 1 - \frac{1}{2\kappa}, \quad \kappa \to \infty .\end{aligned}$$

To get limiting expressions for the right-hand side of (3.7.13), we may, for $m \to 0$, use

$$I_1(sm) = \sum_{r=0}^{\infty} \frac{1}{\Gamma(r+2)\Gamma(r+1)} \left(\frac{sm}{2}\right)^{2r+1},$$

and, for $m \to \infty$, use

$$I_1(sm) \sim (2\pi sm)^{-\frac{1}{2}}\left\{1 - \frac{3}{8sm}\right\} \exp sm .$$

Some calculations then yield the limiting forms of (3.7.13)

$$\kappa \sim m \left(\frac{\pi}{2}\right)^{\frac{1}{2}} , \quad \kappa \to o ,$$

$$\kappa \sim m^2 \quad , \quad \kappa \to \infty . \tag{3.7.14}$$

Criterion (a) suggests we look at the formulae for the densities. von Mises for $\kappa \to o$ is approximately

$$\frac{1}{2\pi} \ (1+\kappa\cos\theta) \ ,$$

while the angular Gaussian is, for $m \to o$

$$\frac{1}{2\pi} \int_0^{\infty} s(1+ms\cos\theta)\exp(-s^2/2)ds,$$

$$= \frac{1}{2\pi} \ (1+ \frac{\sqrt{2\pi}}{2} \ m \ \cos\theta),$$

so we should set $\kappa = m \ (\pi/2)^{\frac{1}{2}}$ as in (3.7.14). When κ is large, θ is trivially shown to be Gaussian with variance κ^{-1}. When $m \to$ large, $\tan\theta = X_2/X_1$ where X_2 is Gaussian (0,1) and X_1 is Gaussian (m,1) and, since θ is small, $\tan\theta = \theta$. Hence θ is asymptotically Gaussian $(0,m^{-2})$. Matching $\kappa = m^2$, again as in (3.7.14). In fact it was found that "drawing" densities on a CRT and matching the AG(m fixed) and VM (κ to be found as $\kappa(m)$) densities by eye gave <u>very</u> similar $\kappa(m)$ to the values got by $E\cos\theta$ matching.

Figure 3.1 shows a sheaf of angular Gaussians. Figure 3.2 shows a sheaf of eye matched ($E\cos\theta$ match would be very similar) von Mises distributions. The relationships may be seen when transparencies of these figures are overlaid. When matched pairs are plotted together, AG exceeds VM at θ=0 and π, and is less in the middle of the range. With Kent's results, mentioned above, we see that the angular Gaussian is the extreme member of the quadruplet -- but the density differences are small. It is in fact hard to appreciate such pictures (see Figures 3.3b, 3.4b, 3.5b) -- perhaps ratios of densities should be used.

Matching method (b) leads us to more revealing pictures of the differences of the cumulatives of the angular Gaussian, von Mises, and Brownian distributions. For the first two distributions, the supremum matched κ is, for all m, slightly greater than the eye -- or $E\cos\theta$ -- matched κ. Of course the suprema obtained are smaller than for these latter methods. As is seen in Figure 3.3a, the supremum is .0125, the same as Stephens' value at $\kappa=1.4$, when the Brownian and von Mises are compared. In Figure 3.3b, the corresponding densities are drawn. Figure 3.4 illustrates the worst situation found (at $\kappa=2.646$, $m=1.6$) for the angular Gaussian and the von Mises -- the supremum is less than in Figure 3.3a. In Figures 3.5, the worst case for Brownian and angular Gaussian is shown.

Turning now to the sphere, Roberts & Ursell did $E\cos\theta$ matching of the Fisher and Brownian distributions (the matching formula being (16) above) and looked also at the matching method (b). They found that the two methods give very similar results. Their conclusion was, of course, that it should be safe to use the Fisher distribution for statistics when the parental distribution was really Brownian.

Figure 3.6 gives a sheaf of Fisher densities for θ, which from (3.7.9) are given by

$$\frac{\kappa}{2\sinh}\exp(\kappa\cos\theta)\sin\theta \tag{3.7.15}$$

for $\kappa = 0(.25)1.5,\ 2(1)10$. Figure 3.7 gives a sheaf of angular Gaussians densities of θ, i.e., plots of

$$\frac{1}{(2\pi)^{\frac{1}{2}}}\int_0^\infty s^2\exp\{-\tfrac{1}{2}(s^2+m^2-2sm\cos\theta)\}ds\ \sin\theta \tag{3.7.16}$$

for $m = 0\ (.2)1,\ 1.25(.25)\ 3$. The matching is vivid only when transparencies of those figures are overlaid. The worst case we encountered is shown in Figure 3.8, where the difference of the cumulatives (Figure 3.8a) and the densities (Figure 3.8b) are shown. The supremum in Figure 3.8a is slightly less than .02, rather than slightly over .02 as Roberts & Ursell found. This occurred for a κ of 4.56, which corresponds to a reasonably dispersed cluster. Our efforts at eye fitting of densities was much less successful here. The supremum fitted kappas are shown in the brief table below:

m	fitted κ
.1	.160
.3	.482
.5	.815
.75	1.26
1	1.74
1.5	2.94
3	9.4
4	16.4
5	25.4

These agree well with the formulae obtained from an $E\cos\theta$ match, or by mathematically matching the density functions (3.7.15) (3.7.16). By calculations similar to those used earlier, we find

$$\kappa \sim 4(2\pi)^{-\frac{1}{2}} m \qquad (\kappa \to 0),$$

$$\kappa \sim m^2 \qquad (\kappa \to \infty). \tag{3.7.17}$$

Thus the angular Gaussian is as good a match as the Brownian to the Fisher distribution, and to the von Mises.

The "matching" formulae limits given in (3.7.14) and (3.7.17) have been generalized by Christopher Bingham (see Appendix C) to q dimensions and read

$$\kappa \sim 2^{\frac{1}{2}} \frac{\Gamma(\frac{q+1}{2})}{\Gamma(\frac{q}{2})} m \qquad (\kappa \to 0)$$

$$\kappa \sim m^2 \qquad (\kappa \to \infty) . \tag{3.7.18}$$

LEGEND

FIGURE 3.1. A sheaf of angular Gaussian densities for $m=0(.2)\ 1.0,\ 1.25\ (.25)\ 2.5$.

FIGURE 3.2 A sheaf of von Mises densities matched by eye to the curves in Figure 3.1. $E\cos\theta$ matchings would be very similar.

FIGURE 3.3. Three methods for assessing the best matching of the Brownian and von Mises distributions when they are hardest to match: $v=1.06$, $\kappa=1.4$.
(a) A plot of the difference BR-VM of the cumulative distributions.
(b) Plots of the br and vm densities.
(c) A plot of the relative error, (BR-VM)/VM .

FIGURE 3.4. Three methods for assessing the best matching of the angular Gaussian and the von Mises distributions when they are hardest to match: $m=1.6$, $\kappa=2.646$.
(a) A plot of the difference AG-VM of the cumulative distributions.
(b) Plots of the ag and vm densities.
(c) A plot of the relative error, (AG-VM)/VM .

FIGURE 3.5. Three methods for assessing the best $E\cos\theta$ matching of the Brownian and angular Gaussian distributions when they are hardest to match: $m=1.2$, $v=0.818$.
(a) A plot of the difference AG-BR of the cumulative distributions.
(b) Plots of the densities ag and br .
(c) A plot of the relative error, (AG-BR)/BR .

FIGURE 3.6. A sheaf of Fisher densities (3.7.15) for θ , for $\kappa=0(.25)\ 1.5,\ 2(1)\ 10$.

FIGURE 3.7. A sheaf of angular Gaussian densities (3.7.16) for θ for $m=0(.2)\ 1,\ 1.25\ (.25)\ 3$.

FIGURE 3.8. Comparisons of the Fisher and angular Gaussian distributions when they are hardest to match: $\kappa=4.56$, $m=2$.
(a) A plot of the difference of the cumulatives for θ , AG-FI .
(b) Plots of the two densities for θ , $ag_3(\theta)$ and $fi(\theta)$.

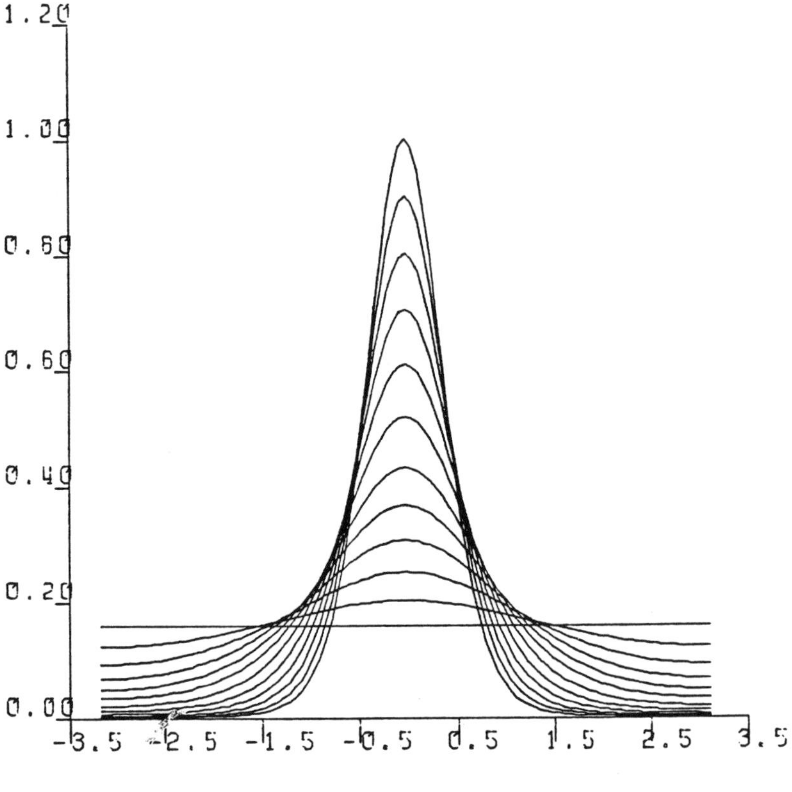

Figure 3.1

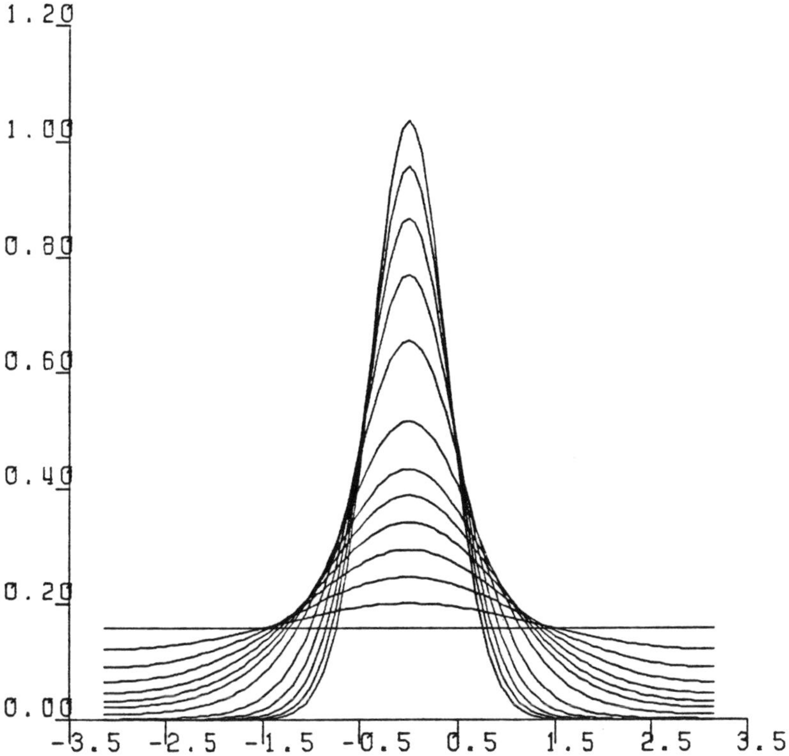

Figure 3.2

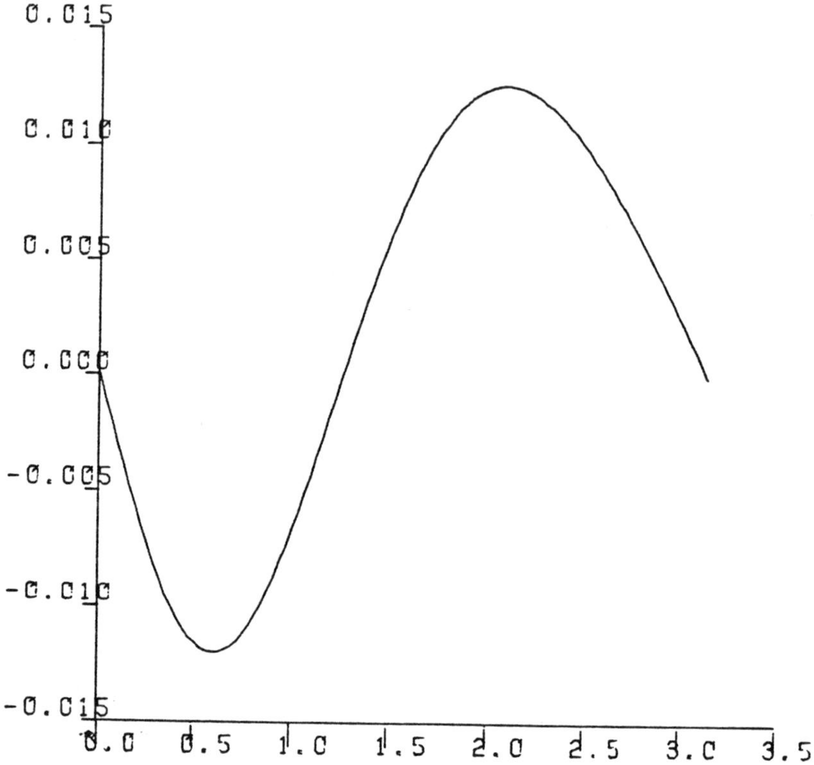

Figure 3.3a

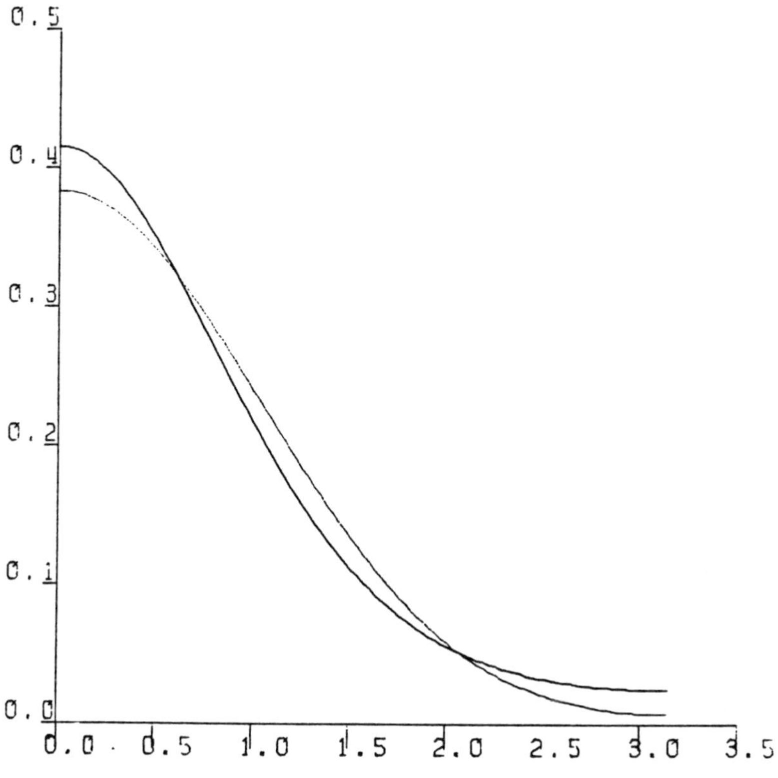

Figure 3.3b

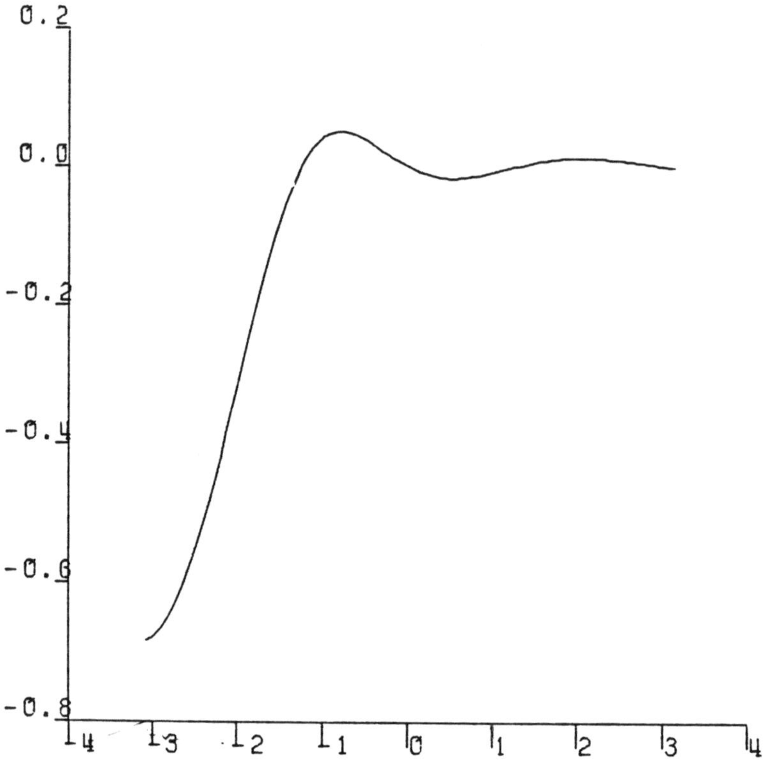

Figure 3.3c

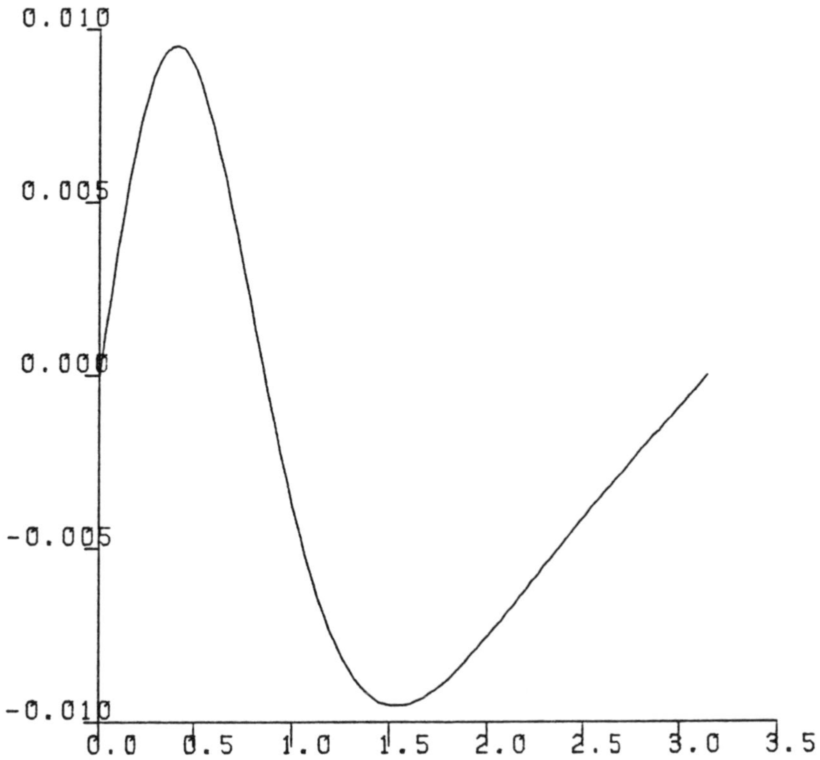

Figure 3.4a

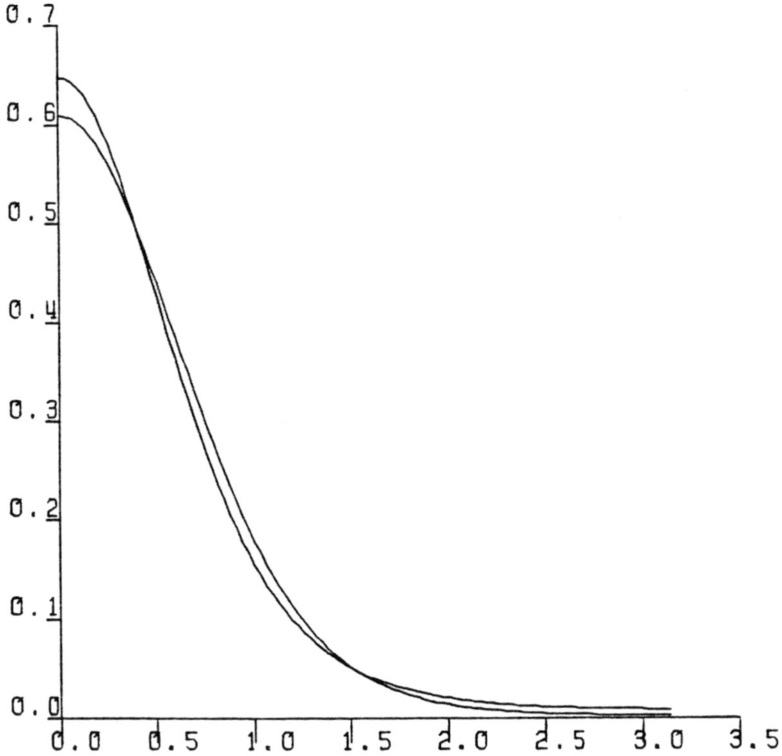

Figure 3.4b

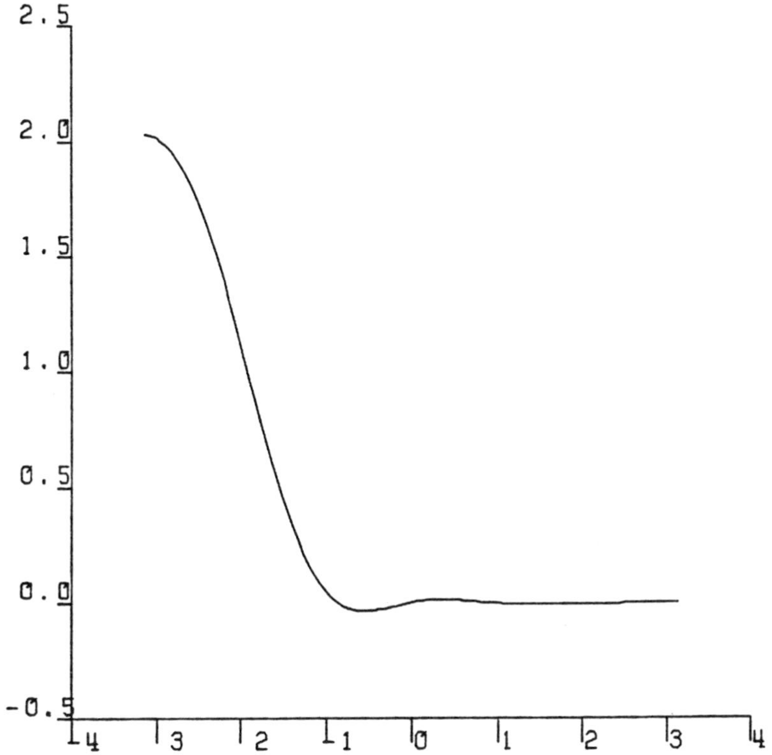

Figure 3.4c

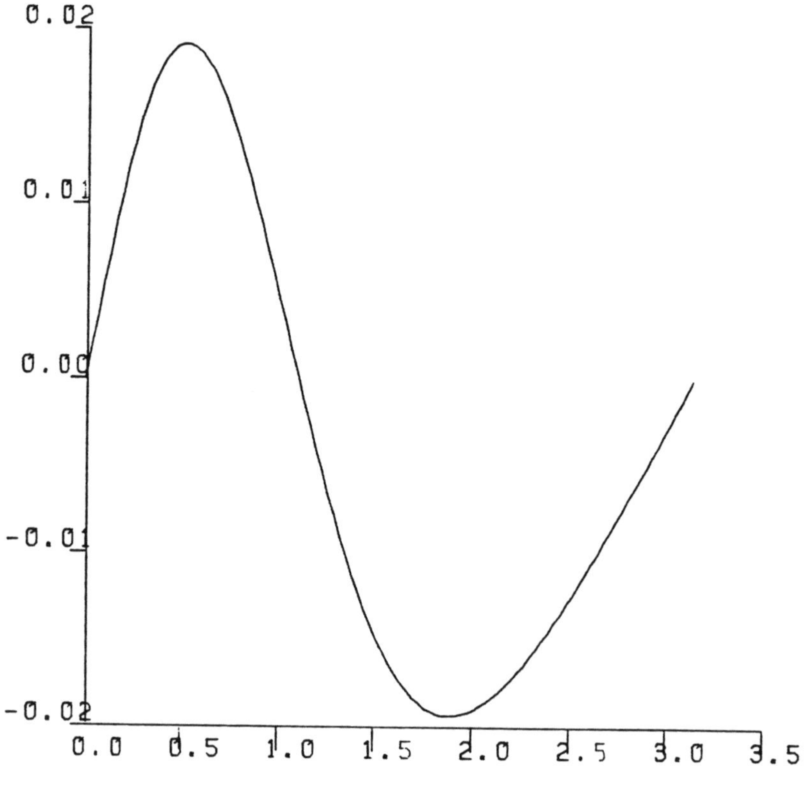

Figure 3.5a

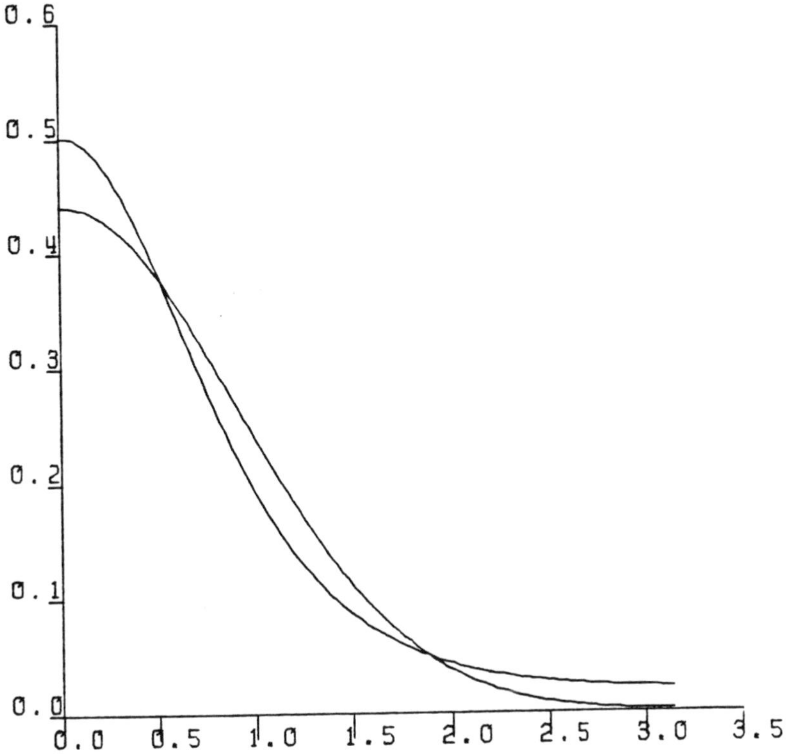

Figure 3.5b

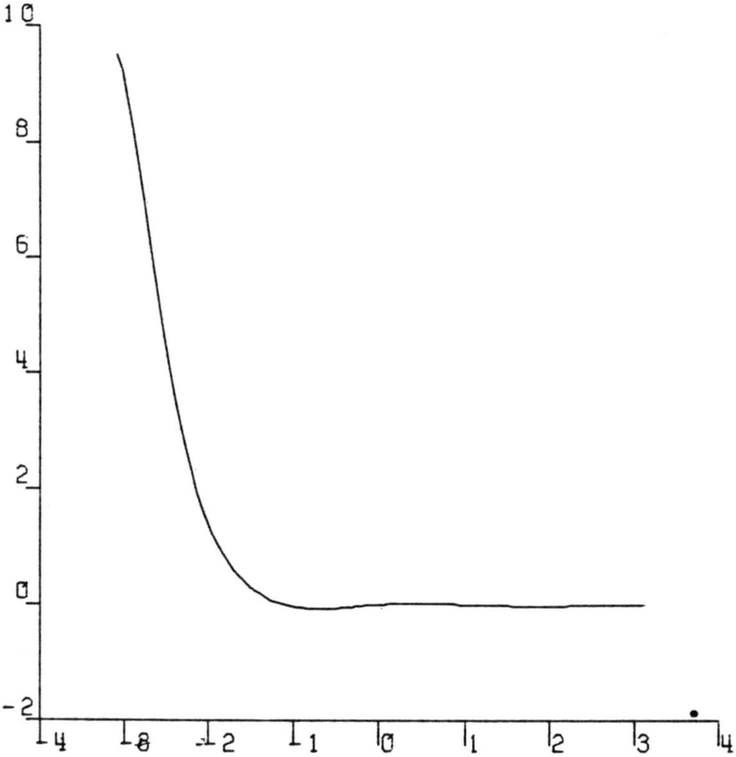

Figure 3.5c

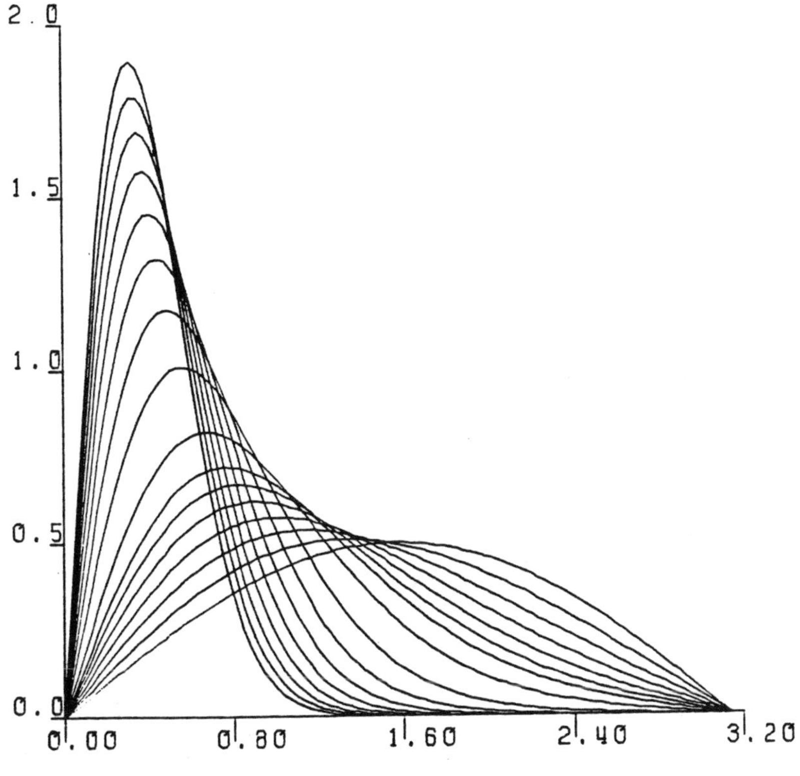

Figure 3.6

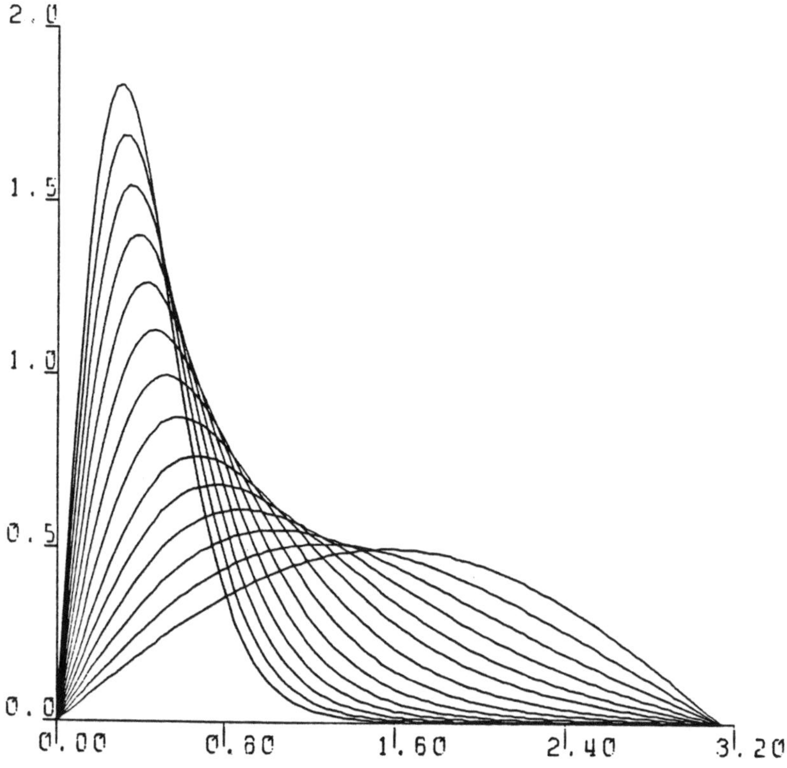

Figure 3.7

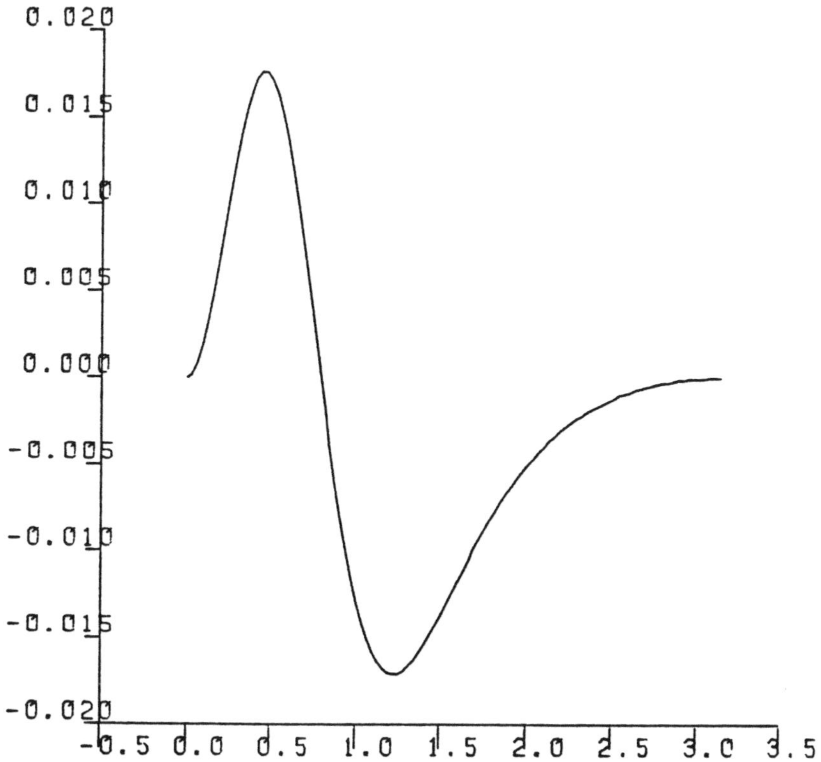
0.020
0.015
0.010
0.005
0.000
-0.005
-0.010
-0.015
-0.020
-0.5
0.0
0.5
1.0
1.5
2.0
2.5
3.0
3.5

Figure 3.8a

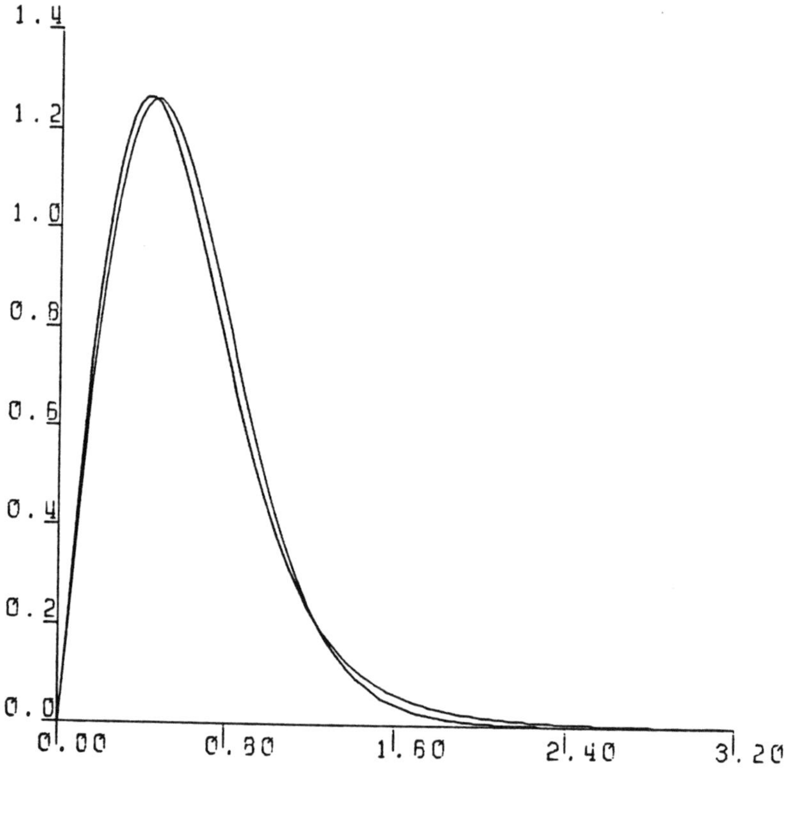

Figure 3.8b

Chapter 4 Statistical Methods Based on the Center of Mass of the Sample

4.1 Introduction

If $x_1, \ldots, x_n$ denote the positions of n unit masses on Ω_q, then their center of mass is at a point within the sphere $\bar{x} = X/n$, $X=x_1+\ldots+x_n$. If the points are clustered near some point μ on Ω_q, then $\bar{x}$ will be near μ. The tighter the cluster, the nearer $\|\bar{x}\|$ will be to its upper bound unity. The direction of $\bar{x}$, $X / \|X\|$, will be close to μ and so if $x_1, \ldots, x_n$ are a random sample from a unimodal distribution on Ω_q, $X / \|X\|$ should be a good estimator of this mode.

If this distribution has the Langevin density proportional to $\exp \kappa \mu' x$ where $\kappa > 0$, the likelihood of the data is proportional to $\exp \kappa \mu' (x_1 + \ldots + x_n)$. Hence the maximum likelihood (m.l.) estimator of μ is $\hat{\mu} = X / \|X\|$. The statistical theory of the Langevin when the sample size is large and (separately), when κ is large, will be developed below.

However the large sample theory of a distribution with density $f(\mu' x)$ is scarcely more difficult to derive. If f is such that $E\mu' x > 0$, it may be expected that $\hat{\mu} = X / \|X\|$ will be a good estimator of μ, at least when n is large. If μ is the mode of $f(\mu' x)$, it should be better still. This approach was first suggested by Saw (1978), and Wellner (1978) in an unpublished paper, and we will give a different and more general development of his ideas below before discussing the Langevin distribution.

4.2 Basic Theory

Let the random vector x on Ω_q have density $f(\mu' x)$ with respect to the invariant measure ω_q, and set $x = t\mu + (1-t^2)^{\frac{1}{2}}\xi$ where ξ is a unit vector orthogonal to μ as in Section 2.2. Then t and ξ are independent, ξ uniform on the unit sphere in $\mu^{\perp}$, and the density of t is, on $(-1, 1)$,

$$\omega_{q-1}\, f(t)(1-t^2)^{\frac{q-3}{2}} \quad ; \tag{4.2.1}$$

we will assume that $f \geqslant 0$ is such that the integral of (4.2.1) is unity, and that

$$Et = \omega_{q-1} \int_{-1}^{1} tf(t)(1-t^2)^{\frac{q-3}{2}}\, dt > 0 \; . \tag{4.2.2}$$

Then

$$Ex = \mu Et \; , \tag{4.2.3}$$

$$M = Exx' = Et^2\mu\mu' + \frac{1-Et^2}{q-1}(I_q - \mu\mu') \; , \tag{4.2.4}$$

$$\Sigma = \text{var}x = \text{var}t\; \mu\mu' + \frac{1-Et^2}{q-1}(I_q - \mu\mu') \; . \tag{4.2.5}$$

All moments of the components of x are finite.

The multivariate central limit theorem (MCLT) applies so that, as $n \to \infty$,

$$\mathcal{L}n^{-\frac{1}{2}}(X-nEt\mu) \to G_q(0, \ddagger) / \tag{4.2.6}$$

Hence, using (2.5),

$$\mathcal{L}n^{-\frac{1}{2}}(\mu'X-nEt) \to G_1(0, \text{vart}) , \tag{4.2.7}$$

$$\mathcal{L}n^{-\frac{1}{2}}X_{\perp\mu} \to G_q(0, \frac{1-Et^2}{q-1}(I_q-\mu\mu')) , \tag{4.2.8}$$

$$\mathcal{L}n^{-1} \| X_{\perp\mu} \|^2 \to \frac{1-Et^2}{q-1} \chi^2_{q-1} , \tag{4.2.9}$$

where $X_{\perp\mu} = (I_q-\mu\mu')X$. It follows that

$$\frac{X}{n} \to \mu Et , \frac{\|X\|}{n} \to Et , \hat{\mu} = \frac{X}{\|X\|} \to \mu \tag{4.2.10}$$

with probability one. These are also consequences of the law of large numbers (LLN).

To describe the accuracy of $\hat{\mu}$ as an estimator of μ , consider

$$\mu'(\mu-\hat{\mu}) = 1-(\mu'X / \|X\|) ,$$

$$= 1-\{1-\| X_{\perp\mu}\|^2 / \|X\|^2\}^{\frac{1}{2}}$$

when $\mu'X \geqslant 0$ because $\|X\|^2 = (\mu'X)^2 + \|X_{\perp\mu}\|^2$.

But when n is large μ', will be positive with probability $\to 1$. By (4.2.9) and (4.2.10), $\|X_{\mu\perp}\| / \|X\|$ will be a random quantity of order n^{-1},

$$\mu'(\mu-\hat{\mu}) \sim \frac{1}{2} \frac{\| X_{\perp\mu}\|^2}{\| X\|^2}$$

$$\sim \frac{1}{2(Et)^2} \frac{\| X_{\perp\mu}\|^2}{n^2} .$$

Thus by (2.9), setting $\mu'\hat{\mu} = \cos\hat{\theta}$,

$$\mathcal{L}\, 2n(1-\cos\theta) \to \frac{1-Et^2}{(Et)^2} \frac{\chi^2_{q-1}}{q-1} . \tag{4.2.11}$$

This enables us to set up a "circular" confidence cone, with axis $\hat{\mu}$ and semi-angle $\hat{\theta}$ which will contain the true μ with a prescribed probability -- or confidence. In practice, we will not know Et and Et^2, but (4.2.11) is still true in the form

$$L \frac{(\hat{Et})^2}{1-(\hat{Et^2})} 2n(1-\cos\theta) \to \frac{\chi^2_{q-1}}{q-1} \tag{4.2.12}$$

if $(\hat{Et})$ and $(\hat{Et^2})$ are consistent estimators of Et and Et^2. For the former, we may use $\hat{Et} = \|X\|/n$ by (4.2.10). For the latter, $n^{-1}\sum_1^n (\hat{\mu}'x_i)^2$ may be used. To prove this, observe that the LLN tells us that $n^{-1}\Sigma(\mu'x_i)^2 \to Et^2$, so we need to show

$$n^{-1}\sum_1^n \{(\hat{\mu}x_i)^2 - (\mu'x_i)^2\} = n^{-1} \Sigma\, (\hat{\mu}+\mu)'x_i(\hat{\mu}-\mu)x_i \tag{4.2.13}$$

tends to zero. By the Cauchy inequality, the modulus of the r.h.s. of (4.2.13) is less than

$$\left[n^{-1} \Sigma \{(\hat{\mu}+\mu)'x_i\}^2 \quad n^{-1} \Sigma \{(\hat{\mu}-\mu)'x_i\}^2 \right]^{\frac{1}{2}} . \qquad (4.2.14)$$

The first factor in (4.2.14) is less than or equal to 4. The second factor can be rewritten as

$$(\hat{\mu}-\mu)'M_n(\hat{\mu}-\mu) \;, \; M_n = n^{-1}\Sigma x_i x_i \;.$$

The LLN guarantees $M_n \to M$, and since $\hat{\mu} \to \mu$, this term tends to zero.

The concentration of the distribution with density $f(\mu'x)$ could be measured by Et, $0 \leqslant Et \leqslant 1$, of which $\|X\|/n$ is the natural estimator. The identity

$$\frac{\|X\|^2}{n^2} - (Et)^2 = \frac{(\mu'X)^2}{n^2} - (Et)^2 + \frac{\|X_{\perp\mu}\|^2}{n^2}$$

yields

$$\left(\frac{\|X\|}{n} + Et\right)\left(\frac{\|X\|}{n} - Et\right) = \left(\frac{\mu'X}{n} + Et\right)\left(\frac{\mu'X}{n}\right) - Et) + \frac{\|X_{\perp\mu}\|^2}{n^2} . \qquad (4.2.15)$$

Multiplying (4.2.15) by $n^{\frac{1}{2}}$, letting $n \to \infty$, and using (4.2.7), (4.2.9) establishes that

$$Ln^{\frac{1}{2}}\left(\frac{\|X\|}{n} - Et\right) \rightarrow Ln^{\frac{1}{2}}\left(\frac{\mu' X}{n} - Et\right) \rightarrow G_1(0, \text{var}t) \ . \tag{4.2.16}$$

Another consequence of (2.15) is that

$$\frac{\|X\|}{n} = \frac{\mu' X}{n} + \frac{1}{2Et}\frac{\|X_{\perp\mu}\|^2}{n^2} + O(n^{-3/2}) \ , \tag{4.2.17}$$

which we will use frequently.

We may use the argument leading to (4.2.11) to establish a significance test of the null hypothesis $\mu = \mu_0$. We will suspect that this null hypothesis is wrong if $\|X_{\perp\mu_0}\|$ is "too large" because X points in the direction of the true μ . But if, in fact, $\mu = \mu_0$, (4.2.9) tells us that

$$\mathcal{L}\ \frac{\|X_{\perp\mu_0}\|^2}{n} \rightarrow \frac{1-Et^2}{q-1}\chi^2_{q-1} \ . \tag{4.2.18}$$

This leads to a computable procedure if we use $n^{-1}\Sigma(\hat{\mu}' x_i)^2 = \hat{\mu}' M_n \hat{\mu}$ instead of the unknown Et^2 . To find the probability of getting a significant result when μ is not μ_0 , the <u>power</u> of the test, we need the distribution of the l.h.s. of (4.2.18) for general μ .

Because $n \rightarrow \infty$, we must set

$$\mu = \frac{\mu_0 + \delta n^{-\frac{1}{2}}}{\|\mu_0 + \delta n^{-\frac{1}{2}}\|} = \mu_0 + \delta_{\perp\mu_0} n^{-\frac{1}{2}} + O(n^{-1}) \ . \tag{4.2.19}$$

and re-examine (4.2.8) which is always true.

$$X_{\perp\mu} = (I_q - \mu\mu')X ,$$

$$\sim \{I_q - \mu_o\mu_o + n^{-\frac{1}{2}}(\mu_o\delta'_{\perp\mu_o} + \delta_{\perp\mu_o}\mu'_o)\}X ,$$

$$X_{\perp\mu} \sim X_{\perp\mu_o} + n^{-\frac{1}{2}}(\mu_o\delta'_{\perp\mu_o}X + \delta_{\perp\mu_o}\mu'_o X) . \tag{4.2.20}$$

Taking expectations of both sides of (4.2.20), we see

$$E\,\frac{X_{\perp\mu_o}}{n^{\frac{1}{2}}} = -\,Et\delta_{\perp\mu_o} + O(n^{-\frac{1}{2}}) . \tag{4.2.21}$$

Since its asymptotic covariance matrix is the limit of that in (4.2.8), we have

$$L n^{-\frac{1}{2}} X_{\perp\mu_o} \to G_q(-Et\delta_{\perp\mu_o} , \frac{1-Et^2}{q-1}(I_q - \mu_o\mu'_o)) \tag{4.2.22}$$

or

$$L\left[\frac{q-1}{n(1-Et^2)}\right]^{\frac{1}{2}} X_{\perp\mu_o} \to G_q(-\frac{Et\delta_{\perp\mu_o}(q-1)^{\frac{1}{2}}}{(1-Et^2)^{\frac{1}{2}}} , I_q - \mu_o\mu'_o) . \tag{4.2.22'}$$

Now if

$$LY = G_q(\eta, P_V)$$

where P_V is an idempotent matrix, rank s, and $P_V\eta = \eta$,

$$L\,\|Y\|^2 = \chi_s^2(\|\eta\|^2),$$

a noncentral chi-square with s degrees of freedom and non-centrality parameter $\|\eta\|^2$.

Thus the distributional statement we seek is

$$L\,\frac{q-1}{1-Et^2}\,\frac{\|X_{\perp\mu_0}\|^2}{n} \to \chi_{q-1}^2\left\{\frac{(Et)^2(q-1)\|\delta_{\perp\mu_0}\|^2}{1-Et^2}\right\}. \quad (4.2.23)$$

Finally we remark that

$$\frac{\|X_{\perp\mu_0}\|^2}{n} = \left(\frac{\|X\|}{n} + \frac{\mu_0'X}{n}\right)(\|X\| - \mu_0'X) \to 2Et(\|X\| - \mu_0'X),$$

so (4.2.23) could also be written as

$$L\,\frac{2(q-1)Et}{1-Et^2}(\|X\| - \mu_0'X) \to \chi_{q-1}^2\left\{\frac{(Et)^2(q-1)\|\delta_{\perp\mu_0}\|^2}{1-Et^2}\right\}. \quad (4.2.24)$$

This gives all the required common procedures for a single sample from the density $f(\mu'x)$. A more general null hypothesis than $\mu=\mu_0$ is $\mu\in V$, where V is some specified subspace of dimension s. A natural test would reject this null hypothesis if $X_{\perp V}$, the part of X orthogonal to V, is too large in magnitude.

Let $P_{V\perp}$ project orthogonally on to $V^\perp$, and suppose that, in fact, $\mu\in V$. Then from (4.2.6), since $X_{\perp V} = P_{V\perp}X$,

$$\mathcal{L}\, n^{-\frac{1}{2}} X_{\perp v} \to G_q\left(0, \frac{1-Et^2}{q-1} P_{v\perp}\right) , \tag{4.2.25}$$

so that

$$\mathcal{L}\, \frac{\| X_{\perp v} \|^2}{n} \to \frac{1-Et^2}{q-1} \chi^2_{q-s} . \tag{4.2.26}$$

To find the power of the test implied by (4.2.26), we must suppose that

$$\mu = \mu_0 + \delta n^{-\frac{1}{2}} , \; \mu_0 \varepsilon V , \; \delta \varepsilon V^{\perp} . \tag{4.2.27}$$

Applying $P_{v\perp}$ to (4.2.6), we find

$$\mathcal{L}\, \frac{\| X_{\perp v} \|^2}{n^{\frac{1}{2}}} \to G\left(\delta Et , \frac{1-Et^2}{q-1} P_{v\perp}\right) ,$$

so that

$$\mathcal{L}\, \frac{\| X_{\perp v} \|^2}{n} \to \frac{1-Et^2}{q-1} \chi^2_{q-s}\left(\frac{(Et)^2 (q-1) \| \delta \|^2}{1-Et^2}\right) , \tag{4.2.28}$$

which generalizes (4.2.23).

4.3 General Several Sample Tests

Given a sample of size n_1 (resultant X_1) from $f_1(\mu_1' x)$ and a sample of size n_2 (resultant X_2) from $f_2(\mu_2' x)$, we might be asked to test whether the two populations had the same concentrations and/or to test whether they have the same axes of rotation. Similar questions arise if $n \geqslant 2$ populations are sampled.

To compare two concentrations, we may use (4.2.16) as follows:

$$\frac{\|X_i\|}{n_i} - (Et)_i \sim G_1\left(0, \frac{(vart)_i}{n_i}\right), \quad i = 1, 2;$$

if $(Et)_1 = (Et)_2$,

$$\frac{\|X_1\|}{n_1} - \frac{\|X_2\|}{n_2} \sim G_1\left(0, \frac{(vart)_1}{n_1} + \frac{(vart)_2}{n_2}\right);$$

$$\left(\frac{(vart)_1}{n_i} + \frac{(vart)_2}{n_2}\right)^{-\frac{1}{2}} \left(\frac{\|X_i\|}{n_1} - \frac{\|X_2\|}{n_2}\right) \sim G(0, 1). \tag{4.3.1}$$

Hence the l.h.s. of (4.3.1), with estimates of $(vart)_i$, would be referred to a table of the standard Gaussian distribution. Similar tactics could be used for more than 2 samples.

Of more interest are the tests for modal or rotational symmetry vectors. With two samples, one might draw a triangle with sides X_1, X_2, and $X_1 + X_2$. Since X_i points in the direction μ_i, roughly, one would doubt the null hypothesis if $\|X_1\| + \|X_2\| - \|X_1 + X_2\|$ is "too large." This however ignores the possible unequal accuracies of X_1 and X_2 as estimators of the directions of μ_1 and μ_2. Let w_1 and w_2 be two positive weights. Then it is equally sensible to use the test statistic

$$w_1\|X_1\| + w_2\|X_2\| - \|w_1X_1 + w_2X_2\|, \tag{4.3.2}$$

particularly if the weights adjust for unequal accuracies. With m samples, the generalization of (4.3.2) is

$$T = \sum_1^m w_i\|X_i\| - \|\sum_1^m w_iX_i\|, \tag{4.3.3}$$

whose asymptotic distribution will now be derived. We will choose the weights in the course of the investigation to give T a simple distribution, and then find that they do reflect the relative accuracies of the estimators.

Suppose first that $\mu_1 = \ldots = \mu_m = \mu$, say, and apply (4.2.17) to each X_i so that

$$\sum_1^m w_i\|X_i\| = \mu' \sum_1^m w_iX_i + \tfrac{1}{2}\sum_1^m \frac{w_i}{(Et)_i} \frac{\|X_{i\perp\mu}\|^2}{n_i} + R. \tag{4.3.4}$$

We want all $n_1, \ldots, n_m$ to tend to ∞, so we could set $n_i = \alpha_iN$, $x_i > 0$, $i=1 \ldots m$. Then R in (4.3.4) is $O(N^{-\frac{1}{2}})$. The version of (4.2.17) appropriate for Σw_iX_i yields

$$\|\sum_1^m w_iX_i\| = \mu' \sum_1^m w_iX_i + \tfrac{1}{2}\frac{\|\Sigma w_iX_{i\perp\mu}\|^2}{\Sigma w_in_i(Et)_i} + O(N^{-\frac{1}{2}}). \tag{4.3.5}$$

Hence

$$2T \backsim \sum_1^m \frac{w_i}{n_i(Et)_i} \| X_{i\perp\mu} \|^2 - \frac{\| \Sigma w_i X_{i\perp\mu} \|^2}{\Sigma w_i n_i (Et)_i} \quad . \tag{4.3.6}$$

Suppose that the w_i are chosen so that

$$L \left\{ \frac{w_i}{n_i(Et)_i} \right\}^{\frac{1}{2}} X_{i\perp\mu} \to G_q(0, I_q - \mu\mu') \;, \tag{4.3.7}$$

and call the random variable with the distribution in the r.h.s. of (4.3.7), Z_i . Inspection of (4.2.8) then shows that

$$w_i = (q-1)(Et)_i \;/\; \{1-(Et^2)_i\} \;. \tag{4.3.8}$$

If we now write

$$\left. \begin{aligned} &\lambda_i = \alpha_i w_i (Et)_i \;/\; \Sigma\alpha_i w_i (Et)_i \;, \\ &\lambda_i > 0, \; \sum_1^m \lambda_i = 1 \;, \end{aligned} \right\} \tag{4.3.9}$$

then

$$2T \backsim \sum_1^m \| Z_i \|^2 - \| \Sigma\lambda_i^{\frac{1}{2}} Z_i \|^2 \quad . \tag{4.3.10}$$

If $\{a_{ij}\}$ is an $m \times m$ orthonormal array with last row $\sqrt{\lambda_1}$, ..., $\sqrt{\lambda_m}$ and we set $Y_i = \Sigma a_{ij} Z_j$, then $Y_1, \ldots, Y_m$ are independent and

To apply the test suggested by (4.3.11), it is necessary to use consistent estimators of the $(Et)_i$, $(Et^2)_i$. To evaluate its power it is also necessary to decide for what $\delta_{i\perp\mu}$ it is required. We will return later to interpret the weights w_i used above.

Another null hypothesis is of interest with m populations. It is that $\mu_1, \mu_2, \ldots, \mu_m$ all belong to some specified subspace V of dimension $q-s$. A natural statistic to use is

$$T' = 2 \sum_1^m u_i \frac{\| X_{i\perp v} \|^2}{n_i} , \tag{4.3.19}$$

where the u_i are non-negative weights to be determined. With experience of the previous calculations, we may immediately seek the distribution of (4.3.19) on the sequence of alternatives

$$\mu_i = \nu_i + \delta_i n_i^{-\frac{1}{2}} , \qquad \nu_i \varepsilon V , \quad \delta_i \varepsilon V_\perp . \tag{4.3.20}$$

Analogously to (4.3.13), we find

$$L n_i^{-\frac{1}{2}} X_{i\perp v} \to G_q\left((Et)_i \delta_i , \frac{1-(Et^2)_i}{q-1} P_{\perp v} \right) , \tag{4.3.21}$$

where $P_{\perp v}$ is the orthogonal projector onto $V^\perp$. In order that the asymptotic distribution of the statistic (4.3.19) be a chi-square, we must choose

$$u_i = (q-1) / \{1 - (Et^2)_i\} \cdot \quad (4.3.22)$$

Then

$$LT' \to \chi^2_{sm} \left(\sum_1^m u_i (Et)_i^2 \, \|\delta_i\|^2\right) . \quad (4.3.23)$$

A comparison of (4.3.17), (4.3.18), and (4.3.23) shows that the components of the non-centrality parameter are the same,

$$\|\delta\|^2 (q-1)((Et)_i)^2 / 1-(Et^2)_i \quad , \quad (4.3.24)$$

when the formulae of w_i and u_i are inserted. Now it should be easier to pick up a deviation from the null hypothesis of size $\|\delta\|$ if both $(Et)_i$ and $(Et^2)_i$ are large, because in those two respects the i th population is concentrated about its axis of rotation; (4.3.24) shows that this is so. Further support for our choice of w_i and u_i comes in the next section.

4.4 The Langevin Distribution with Large Samples

The density of the Langevin distribution in Ω_q can be written

$$a_q(\kappa)^{-1}\exp\kappa\mu'x \qquad (\kappa \geqslant 0) \ , \tag{4.4.1}$$

where

$$\left.\begin{aligned} a_q(\kappa) &= (2\pi)^{q/2} I_{(q/2)-1}(\kappa)\kappa^{-(q/2)+1} \\ &= \omega_q \, {}_0F_1(q/2 \ , \ (\kappa/2)^2) \end{aligned}\right\} , \tag{4.4.2}$$

where $I_\nu(\cdot)$ is the modified Bessel function of the first kind and order ν - see, e.g., Abramowitz and Stegun (ibid, Section 9.6.). The distribution is rotationally symmetric about μ with a single mode at $x=\mu$. As κ increases from zero to infinity, it changes from the uniform distribution to a distribution concentrated wholly at $x=\mu$.

It is a simple member of the exponential family, and its statistical theory will depend upon the properties of

$$A_q(\kappa) = \frac{d}{d\kappa}\log a_q(\kappa) \quad . \tag{4.4.3}$$

While these may be derived from (4.4.2) and the properties of Bessel functions, it is usually easier to work directly with the definition

$$a_q(\kappa) = \int_\Omega \exp\kappa\mu'x \, d\omega_q \quad , \tag{4.4.4}$$

which is studied in Appendix A.

Thus, e.g., we find that A satisfies the Ricatti equation

$$1 - A_q'(u) - A_q^2(\kappa) = (q-1)A_q(\kappa)/\kappa \ , \tag{4.4.5}$$

that

$$A_q(\kappa) = \frac{\kappa}{q} - \frac{\kappa^3}{q^2(q+2)} + \frac{2\kappa^5}{q^3(q+2)(q+4)} + O(\kappa^7) \ , \tag{4.4.6}$$

and

$$A_q(\kappa) = 1 - \frac{q-1}{2}\frac{1}{\kappa} + \frac{(q-1)(q-3)}{8}\frac{1}{\kappa^2} + \frac{(q-1)(q-3)}{8}\frac{1}{\kappa^3} + O(\kappa^{-4}) \ . \tag{4.4.7}$$

Further it may be shown that $A_q(\kappa)$ increases steadily from zero to unity as κ goes from 0 to ∞, and is concave. It is easy to see that $(t=\mu'x)$

$$Et = A_q(\kappa) \ , \ \text{var}t = A_q'(\kappa) \ , \ Et^2 = 1 - \frac{q-1}{\kappa} A_q(\kappa) \ , \tag{4.4.8}$$

$$\begin{aligned} \Sigma &= Exx' - (Ex)(Ex)' \ , \\ &= A_q' \mu\mu' + \frac{A_q}{\kappa}(I_q - \mu\mu') \ . \end{aligned} \tag{4.4.9}$$

Thus, the asymptotic theory will be based upon

$$Ln^{-\frac{1}{2}}(X-nEt\mu) \to G_q(0,\Sigma) \ . \tag{4.4.10}$$

This should be viewed as a special case of (4.2.6) which says

$$Ln^{-\frac{1}{2}}(x-nEt\mu) \to G_q(0, \Sigma) \ , \tag{4.4.11}$$

where Σ is the generalization of (4.4.9):

$$\Sigma = \mathrm{var}t\mu\mu' + \frac{1-Et^2}{q-1}(I_q-\mu\mu') \ .$$

To use the formulae in the last two sections, we need only to replace

$$\left.\begin{array}{l} Et \text{ by } A_q(\kappa) \ , \\ \mathrm{var}t \text{ by } A_q' \ , \\ \dfrac{1-Et^2}{q-1} \text{ by } \dfrac{A_q}{\kappa} \ . \end{array}\right\} \tag{4.4.12}$$

For the Langevin distribution, $\hat{\mu} = X / \|X\|$ is the m.l. estimator of μ and the m.l. estimator of κ, $\hat{\kappa}$ satisfies

$$A_q(\hat{\kappa}) = \frac{\|X\|}{n} \ , \tag{4.4.13}$$

an equation with a unique solution, and leads to a consistent estimator of κ.

It follows immediately that

$$Ln^{\frac{1}{2}}A_q'(\kappa)\ (\hat{\kappa}-\kappa)\ \rightarrow\ G_1(0,\ A_q'(\kappa))$$

by using (4.2.16) and 4.4.12) twice, so that

$$L\ (nA_q(\kappa))^{\frac{1}{2}}(\hat{\kappa}-\kappa)\ \rightarrow\ G_1(0,\ 1)\ . \tag{4.4.14}$$

If we set

$$u(\kappa)\ =\ \int^{\kappa}\{A_q'(\kappa)\}^{\frac{1}{2}}dk\ , \tag{4.4.15}$$

in which the lower terminal is any convenient constant (e.g. zero) then

$$Ln^{\frac{1}{2}}(u(\hat{\kappa})\ -\ u(\kappa))\ \rightarrow\ G_1(0,\ 1) \qquad 4.4.16$$

The transform $u(\hat{\kappa})$ also has less skewness than $\hat{\kappa}$, so (4.4.16) is an excellent basis for making interferences on κ. Further, a circular confidence cone for μ about $\hat{\mu}$ with semi-angle $\hat{\theta}$ follows from (4.2.11), which here reads

$$L2n\ \kappa\ A_q(\kappa)(1-\cos\hat{\theta})\ \rightarrow\ \chi^2_{q-1}\ . \tag{4.4.17}$$

To use this in practice, κ will be replaced by $\hat{\kappa}$ - we no longer need to use estimates of Et and Et^2 .

The tests suggested in the previous sections are easily rewritten. We just list the results.

H: $\mu \varepsilon V$ given one sample size n (c.f. (4.2.26), (4.2.28))

$$L \frac{\| X_{\perp V} \|^2}{n} \to \frac{A_q(\kappa)}{\kappa} \chi^2_{q-s} (\kappa A_q(\kappa) \| \delta \|^2) \tag{4.4.18}$$

Here K: $\mu = \mu_0 + \delta n^{-\frac{1}{2}}$, $\mu_0 \varepsilon V$, $\delta \varepsilon V^{\perp}$,

and s = dimension of V.

H: $\mu_i = \mu$ (unknown), samples size n_i ($i=1, \ldots, m$) (c.f. (4.3.3), (4.3.17), (4.3.18))

$$L 2\left(\sum_1^m \kappa_i \| X_i \| - \| \sum_1^m \kappa_i X_i \| \right) \to \chi^2_{(m-1)(q-1)}(\lambda) , \tag{4.4.19}$$

$$\lambda = \Sigma \kappa_i A_q(\kappa_i) \| \delta_{i\perp\mu} \|^2 - \frac{\| \Sigma \alpha_i \kappa_i A_q(\kappa_i)^{\frac{1}{2}} \delta_{i\perp\mu} \|^2}{\Sigma \alpha_i \kappa_i A_q(\kappa_i)} , \tag{4.4.20}$$

where $n_i = \alpha_i N$, $i=1, \ldots, m$, and

$$K: \mu_i = \mu + \delta_{i\perp\mu} n_i^{-\frac{1}{2}} .$$

Remarks: $w_i = \kappa_i$ and the statistic is the likelihood ratio statistic for the test of H versus K, κ_i's known.

H: $\mu_i \varepsilon V$ (dimension q-s), sample sizes n_i, i=1, ..., n

(c.f. (4.3.19), (4.3.20), (4.3.22), (4.323))

$$L2 \sum_{i=1}^{m} \frac{\kappa_i}{nA_q(\kappa_i)} \| X_{i \perp V} \| \to L2 \sum_{i=1}^{m} \kappa_i (\| X_i \| - \| X_{iV} \|) \to \chi^2_{sm} \left(\sum_{i=1}^{m} \kappa_i A_q(\kappa_i) \| \delta_i \|^2 \right) \tag{4.4.21}$$

where

$$K: \ \mu_i = \nu_i + \delta_i n_i^{-\frac{1}{2}} , \ \nu_i \varepsilon V , \ \delta_i \varepsilon V^{\perp} .$$

Remark: The statistic is the likelihood ratio of statistic for testing H versus K , κ_i's known.

To use the tests (4.4.18), (4.4.19), (4.4.21) in practice, the κ_i will be replaced by their consistent estimators $\hat{\kappa}_i$.

4.5 Concentrated Langevin Distributions

As $\kappa \to \infty$, the Langevin distribution (4.4.1) concentrates about μ because $\exp \kappa \mu' x$ falls off rapidly as x goes away from μ . Setting $x = t\mu + (1-t^2)^{\frac{1}{2}}\xi$ as in Section 4.2, the density of t is, on $(-1, 1)$,

$$\frac{\omega_{q-1}}{a_q(\kappa)} \exp\kappa t \, (1-t^2)^{(q-3)/2} . \tag{4.5.1}$$

When $\kappa \to \infty$, the formula (4.4.2) for $a_q(\kappa)$ may be simplified because

$$I_\nu(\kappa) = (2\pi\kappa)^{\frac{1}{2}}\exp\kappa \, \{1+O(\kappa^{-1})\} . \tag{4.5.2}$$

Thus if we set $u = 2\kappa(1-t)$, it follows that

$$u = 2\kappa(1-t) \underset{d}{\to} U , \; \mathcal{L}U = \chi^2_{q-1} . \tag{4.5.3}$$

In particular,

$$t = \mu' x \to 1 \text{ (in prob) as } \kappa \to \infty . \tag{4.5.4}$$

Since $x_{\perp\mu} = (1-t^2)^{\frac{1}{2}}\xi$,

$$\kappa^{\frac{1}{2}} x_{\perp\mu} = \{\kappa(1+t)(1-t)\}^{\frac{1}{2}}\xi \underset{d}{\to} U^{\frac{1}{2}}\xi , \tag{4.5.5}$$

where U and ξ are independent. By a characterization of the Gaussian, it follows that

$$\kappa^{\frac{1}{2}} X_{\perp\mu} \underset{d}{\rightarrow} z \ , \ Lz = G_q(0, P_{\perp\mu}) \ , \tag{4.5.6}$$

where $P_{\perp\mu} = I_q - \mu\mu'$.

We may now explore the implications of these results for statistical procedures using the resultant X , as in Watson (1981). Since $\mu' X = \Sigma \mu' x_i$, $X_{\perp\mu} = \Sigma x_{i\perp\mu}$, it follows that, as $\kappa \rightarrow \infty$,

$$L \ 2\kappa(n - \mu' X) \rightarrow \chi^2_{n(q-1)} \ , \tag{4.5.7}$$

$$\frac{\kappa^{\frac{1}{2}}}{n} X_{\perp\mu} \underset{d}{\rightarrow} Z \ , \ LZ = G_q(0, P_{\perp\mu}) \ , \tag{4.5.8}$$

$$L \ \kappa \frac{\| X_{\perp\mu} \|^2}{n} \rightarrow \chi^2_{q-1} \ . \tag{4.5.9}$$

Trivial consequences are

$$\mu' X \rightarrow n \ , \ \| X_{\perp\mu} \| \rightarrow 0 \ , \ \| X \| \rightarrow n \ (\text{in prob}) \ . \tag{4.5.10}$$

The program is to examine the statistics and hypothesis referred to above in (4.4.18), (4.4.19), and (4.4.21), when the sample sizes are fixed but the κ_i are large. From (4.4.7), we see that, as $\kappa \rightarrow \infty$,

$$A_q(\kappa) \rightarrow 1 \ , \ A_q'(\kappa) \rightarrow 0 \ . \tag{4.5.11}$$

Here alternative hypotheses must converge to the null hypotheses at the rate $\kappa^{-\frac{1}{2}}$ instead of $n^{-\frac{1}{2}}$ used in the last section.

To test $H: \mu \varepsilon V$, dim $V=S$, given a sample of size n, we should look at $X_{\perp V} = P_{\perp V}X$. If H is true, $X_{\perp V} = P_{\perp V}X_{\perp\mu}$, so by (4.5.8),

$$L\kappa^{\frac{1}{2}}X_{\perp V} \to G(0,\ nP_{\perp V}) \ . \tag{4.5.12}$$

Hence we obtain the analogue of (4.4.18):

$$L\ \kappa\frac{\|X_{\perp V}\|^2}{n} \to \chi^2_{q-s} \ . \tag{4.5.13}$$

If the alternative is

$$K\colon \mu = \mu_0 + \delta\kappa^{-\frac{1}{2}}\ ,\ \mu_0 \varepsilon V\ ,\ \delta\varepsilon V^{\perp}\ ,$$

then $x = t\mu + (1-t^2)^{\frac{1}{2}}\xi_\mu$, $t = \mu' x$, $\xi_\mu \perp \mu$ yields $x_{\perp V} = tP_{\perp V}(\mu_0+\delta\kappa^{-\frac{1}{2}}) + (1-t^2)^{\frac{1}{2}}P_{\perp V}\xi_\mu$. But as $\kappa \to \infty, t \to 1$ so that, since $P_{\perp V}\mu_0 = 0$, $\xi_\mu \to \xi$, we now have

$$\kappa^{\frac{1}{2}}x_{\perp V} = \delta + \{2\kappa(1-t)\}^{\frac{1}{2}}\xi \ .$$

Hence instead of (4.5.6), we have

$$L\ \kappa^{\frac{1}{2}}x_{\perp V} \to G_q(\delta,\ P_{\perp V}) \ . \tag{4.5.14}$$

It follows that, as $\kappa \to \infty$,

$$L\ \kappa \frac{\|X_{\perp v}\|^2}{n} \to \chi^2_{q-s}(\|\delta\|^2)\ . \tag{4.5.15}$$

Since

$$\|X\|^2 - \|X_v\|^2 = \|X_{\perp u}\|^2\ , \tag{4.5.16}$$

and $\|X\|$, $\|X_v\| \to n$ as $\kappa \to \infty$, (4.5.15) may also be written

$$L\ 2\kappa(\|X\| - \|X_v\|) \to \chi^2_{q-s}\ (\|\delta\|^2)\ .$$

To test the hypothesis H that $\mu_1 = \ldots = \mu_m$ where $\kappa_1, \ldots, \kappa_n$ are known and large we can again consider

$$2(\Sigma\kappa_i\|X_i\| - \|\Sigma\kappa_i X_i\|)\ , \tag{4.5.17}$$

but with the alternatives

$$K:\ \mu_i = \mu + \delta_i\kappa_i^{-\frac{1}{2}}\ ,\quad \delta_i \perp \mu\ ,\ i = 1, \ldots, m\ . \tag{4.5.18}$$

The statistic (4.5.17) may be rewritten with the aid of the geometric formula

$$\|a+\delta\| = 1 + a'\delta + \tfrac{1}{2}\|\delta_{\perp a}\|^2 + O(\|\delta\|^3)\ ,\ \|a\| = 1\ ,\ \|\delta\| \text{ small.} \tag{4.5.19}$$

From this formula, it is easy to discuss (4.5.17).

We find that (4.5.17) is asymptotically distributed as

$$\Sigma\kappa_i \frac{\|X_{i\perp\mu}\|^2}{n_i} - \frac{\|\Sigma\kappa_i X_{i\perp\mu}\|^2}{\Sigma\kappa_i n_i} \quad . \tag{4.5.20}$$

Using (4.5.14) for each population, we have

$$\left(\frac{\kappa_i}{n_i}\right)^{\frac{1}{2}} X_{i\perp\mu} \to Z_i \ , \tag{4.5.21}$$

$$L\ Z_i = G_q(n_i^{\frac{1}{2}}\delta_i \ , \ P_\mu) \ .$$

Assuming the κ_i's go to infinity in such a way that

$$\lambda_i^* = \lim_{\kappa_i\text{'s}\to\infty} \left(\frac{\kappa_i n_i}{\sum_1^m \kappa_i n_i}\right) \tag{4.5.22}$$

exist, we see that

$$2(\Sigma\kappa_i\|X_i\| - \|\Sigma\kappa_i X_i\|) \underset{d}{\to} \Sigma\|Z_i\|^2 - \|\Sigma\lambda_i^{\frac{1}{2}} Z_i\|^2 \ . \tag{4.5.23}$$

We may now use the methods used to study (4.3.10) to show that

$$L\ 2(\Sigma\kappa_i\|X_i\| - \|\Sigma\kappa_i X_i\|) \to \chi^2_{(m-1)(q-1)}(\lambda), \tag{4.5.24}$$

where

$$\lambda = \Sigma n_i \| \delta_i \|^2 - \| \Sigma (\lambda_i^* n_i)^{\frac{1}{2}} \delta_i \|^2 . \tag{4.5.25}$$

These are the analogues of (4.4.19) and (4.4.20).

Finally, to test H that $\mu_i \varepsilon V (i = 1, \ldots, m)$ against

$$K: \mu_i = \mu_{iV} + \delta_i \kappa_i^{-\frac{1}{2}} , \mu_{iV} \varepsilon V , \delta_i \varepsilon V^{\perp} \ (i = 1, \ldots, m) , \tag{4.5.26}$$

the proposed statistic is either of the asymptotically equivalent forms

$$\Sigma \kappa_i \frac{\| X_{i \perp V} \|^2}{n} \text{ or } 2 \Sigma \kappa_i (\| X_i \| - \| X_{iV} \|) . \tag{4.5.27}$$

Similar methods show that the limiting distribution of (4.5.27), given (4.5.26), is

$$\chi^2_{sm} (\Sigma_{ni} \| \delta_i \|^2) , \tag{4.5.28}$$

the analogue of (4.4.21).

However these results are not usually of practical importance because, unlike the last section, we do not have consistent estimators to substitute for the unknown κ_i's .

If however,

$$\kappa_1 = \ldots = \kappa_m = \kappa , \kappa \to \infty , \tag{4.5.29}$$

very useful procedures are available which have their origin in Watson (1956), Watson and Williams (1956).

With data from one Langevin population, the m.l. estimators of κ when μ is known, and when it is not, satisfy the equations

$$A_q(\hat{\kappa}_\mu) = \frac{\mu' X}{n}, \quad A_q(\hat{\kappa}) = \frac{\|X\|}{n}. \tag{4.5.30}$$

By (4.4.7), if $\hat{\kappa}_\mu$ and $\hat{\kappa}$ are large, we may write

$$\hat{\kappa}_\mu = \frac{q-1}{2}\,\frac{n}{n-\mu' X}, \quad \hat{\kappa} = \frac{q-1}{2}\,\frac{n}{n-\|X\|}. \tag{4.5.31}$$

Since κ is a concentration parameter, its reciprocal is a dispersion parameter. Hence we could interpret the terms in the identity

$$n - \mu' X = (n - \|X\|) + (\|X\| - \mu' X) \tag{4.5.32}$$

as

$$\left\{\begin{array}{c}\text{dispersion of sample}\\ \text{about true mode}\end{array}\right\} = \left\{\begin{array}{c}\text{dispersion of sample}\\ \text{about sample mode}\end{array}\right\} + \left\{\begin{array}{c}\text{dispersion of}\\ \text{sample mode}\\ \text{about true mode}\end{array}\right\}. \tag{4.5.33}$$

For readers familiar with analysis of variance, the formula

$$\Sigma(x_i-\mu)^2 = \Sigma(x_i-\bar{x})^2 + n(\bar{x}-\mu)^2 \tag{4.5.34}$$

has a similar sense, and they will know that

$$F_{1,n-1} = t^2 = \frac{n(\bar{x}-\mu)^2}{\Sigma(x_i-\bar{x})^2/(n-1)} \tag{4.5.35}$$

is used to test that the sample came from a Gaussian population with mean μ . t has a nice distribution because the terms on the r.h.s. of (4.5.34) are independent and because the first is distributed as $\sigma^2\chi^2_{n-1}$, the second as $\sigma^2\chi^2_1$. The unknown value σ^2 cancels in the ratio.

It is shown in Watson (1981) that, as $\kappa \to \infty$,

$$2\kappa(n-\mu'X) = 2\kappa(n-\|X\|) + 2\kappa(\|X\| - \mu'X) \tag{4.5.36}$$

corresponds to the distributional statement

$$\chi^2_{n(q-1)} = \chi^2_{(n-1)(q-1)} + \chi^2_{q-1} \ . \tag{4.5.37}$$

Hence

$$L\,\frac{(\|X\| - \mu_0' X)/(q-1)}{(n-\|X\|)/(n-1)(q-1)} \to F_{q-1,(q-1)(n-1)} \tag{4.5.38}$$

as $\kappa \to \infty$. So we have a simple test of $H: \mu = \mu_0$. When $K: \mu = \mu_0 + \delta\kappa^{-\frac{1}{2}}$, the l.h.s. of (4.5.38) has a limiting non-central F distribution.

By similar reasoning, to test the equality of the modal vectors of m populations (with the same, but large, concentrations), we may use

$$L \frac{(\Sigma \| X_i \| - \| \Sigma X_i \|)/(m-1)(q-1)}{(n - \Sigma \| X_i \|)/(n-m)(q-1)} \to F_{(m-1)(q-1),(n-m)(q-1)} , \qquad (4.5.39)$$

where $n = \Sigma n_i$.

To test that the means of m populations lie in an s dimensional subspace V , we may use the fact that

$$L \frac{\Sigma \| X_i \| - \| X_{iV} \|)/ms}{(n - \Sigma \| X_i \|)/(n-m)(q-1)} \to F_{ms,(n-m)(q-1)} . \qquad (4.5.40)$$

Chapter 5 Statistical Methods Based on the Moment of Inertia of the Sample

5.1. Introduction

In the last chapter, we developed statistical methods based upon the center of mass $\bar{x} = (x_1+\ldots+x_n)/n$ of unit masses placed at the n sample points $x_1,\ldots,x_n$. As explained in Section 1.3, the moment of inertia of these masses about an axis a is given by $n - a'(\sum_1^n x_i x_i')a$ and will, for varying a, tell us something about the distribution of the data. Thus we were led to consider the eigenvectors and values

$$M_n = n^{-1} \sum_1^n x_i x_i' \quad . \tag{5.1.1}$$

If the data is a random sample from some distribution on Ω_q, then

$$M_n \longrightarrow M = Exx' \quad \text{(in prob.)}. \tag{5.1.2}$$

In general we may write M in spectral form:

$$M = \sum_1^r \lambda_j P_j \quad (r \leq q), \tag{5.1.3}$$

where $\lambda_1,\ldots,\lambda_r > 0$, and the P_j are orthogonal projectors so that $P_j P_k = \delta_{jk} P_j$, $P_j' = P_j$. Let $q_j = \text{rank } P_j$ so that $\sum_1^r q_j \leqslant q$. The λ_j are the eigenvalues of M, which has unit trace so that $\sum_1^r q_j \lambda_j = 1$. If $\sum q_j < q$, M has a zero eigenvalue with multiplicity $q - \sum q_i$ and x is, with probability one, orthogonal to a subspace of this dimension. In practical situations, usually $\sum q_j = q$ so that $\sum_1^r P_j = I_q$ and we will assume this.

For example, if the distribution of x is rotationally symmetric around a unit vector μ, then M must have the form (see 4.2.4)

$$M = Et^2 \mu\mu' + \frac{1-Et^2}{q-1} (I_q - \mu\mu'), \tag{5.1.4}$$

where $t = \mu' x$. Thus M is singular only if $Et^2 = 1$ or 0, that is $x = \pm \mu$ (prob. 1) or $x \perp \mu$ (prob. 1). If in this case $Ex = c\mu$, $c > 0$, we have, in Chapter 4, explained how $\bar{x}$ may be used to estimate μ. But provided Et^2 does not equal $(1-Et^2)(q-1)^{-1}$, the eigenvectors of M_n also provide information about μ. If however $Ex = 0$, $\bar{x}$ is useless, but M_n will still be helpful provided $Et^2 \neq 1/q$.

A simple example with $Ex = 0$ is the Scheiddegger-Watson distribution with a density proportional to $\exp \kappa (\mu' x)^2$. In this case, the log-likelihood of the data is a constant plus $\kappa n \mu' M_n \mu$. Hence if $\kappa > 0$, the m.ℓ. estimator $\hat{\mu}$ of μ is the eigenvector of M_n corresponding to its largest eigenvalue $\lambda_1(M_n)$. If $\kappa < 0$, $\hat{\mu}$ is the eigenvector corresponding to the least eigenvalue $\lambda_q(M_n)$ of M_n.

Another example is a distribution with density on Ω_q,

$$c^{-1}(K) f(x' K x) , \qquad (5.1.5)$$

where the spectral decomposition of K is

$$K = \sum_1^r \kappa_i Q_i ,$$

where the Q_i are orthogonal projectors, $\sum_1^r Q_i = I_q$. Thus

$$x = \left(\sum_1^r Q_i\right) x = Q_1 x + \dots + Q_r x,$$

$$= y_1 + \dots + y_r ,$$

so that

$$Exx' = c(K)^{-1} \int_{\Omega_q} \sum_{i,j} y_i y_j' \, f\left(\sum \kappa_i \|y_i\|^2\right) d\,\omega_q .$$

Since the distribution is unchanged by the transformation y_i to $-y_i$, it follows that $E\, y_i y_j' = 0$, $i \neq j$, and hence

$$M = Exx' = Ey_1y_1' + \ldots + Ey_ry_r' = \lambda_1 Q_1 + \ldots + \lambda_r Q_r , \tag{5.1.6}$$

so that $P_j = Q_j$. Thus the eigenvectors of M_n should help us estimate the Q_j's. A special case of this is the Bingham distribution

$$c(K)^{-1} \exp x' K x , \tag{5.1.7}$$

where we see, because $x'x = 1$, that the eigenvalues of K can only be determined to an additive constant. Bingham (1974) gives an extensive theory of (5.1.7) for $q = 3$. A further example is given in (3.6.7).

Another example which will be dealt with in detail later is the generalized Scheiddegger-Watson distribution with density

$$f(\|x_V\|) , \tag{5.1.8}$$

where x_V is the part of x in an s-dimensional subspace V. If P_V and $P_{V\perp}$ are the orthogonal projections onto V and $V^{\perp}$ respectively, then $Ex = 0$ and

$$M = \frac{E\|x_V\|^2}{s} P_V + \frac{1-E\|x_V\|^2}{q-s} P_{V\perp}. \tag{5.1.9}$$

If $E\,\|x_V\|^2 \neq s/q$, the two distinct eigenvalues of M are different, and M_n may be used to estimate V.

The above discussions presume that we can make the correct correspondences between the eigenvalues and vectors of M_n and those of M. Now unless the distribution of x is restricted to a subspace of $\mathbb{R}^q$, the eigenvalues of M_n will all be distinct with probability one - see, e.g., Okamoto (1973). In the above examples, the eigenvalues of M are usually not all distinct. Furthermore, the distribution theory of eigenvalues and eigenvectors is, for finite n, difficult, if not impossible, to obtain. Both these problems will be greatly eased if we restrict ourselves to large sample theory.

The multivariate central limit theorem (MCLT) holds for independent identically distributed observations on Ω_q so that

$$n^{\frac{1}{2}}(M_n - M) \xrightarrow[d]{} G, \qquad (5.1.10)$$

where G is a $q \times q$ symmetric matrix whose $q(q+1)/2$ functionally independent entries are jointly Gaussian with zero means and a covariance matrix which can be determined. If the elements of x are $x^1, \ldots, x^q$, then

$$\begin{aligned} \operatorname{cov}(G_{ij}, G_{k\ell}) &= \operatorname{cov}(x^i x^j, x^k x^\ell) \\ &= Ex^i x^j x^k x^\ell - Ex^i x^j Ex^k x^\ell . \end{aligned}$$

From (5.1.10) and the identity

$$M_n = M + n^{-\frac{1}{2}}\{n^{\frac{1}{2}}(M_n - M)\},$$

we will be able to write, as $n \longrightarrow \infty$,

$$M_n = M + n^{-\frac{1}{2}} G, \qquad (5.1.11)$$

which shows M_n as a small linear Gaussian perturbation of M. Using the results in Appendix B, we have, when M is defined by (5.1.3) and below, the following are facts about M_n as $n \longrightarrow \infty$. The eigenvalues of M_n form r clusters, of sizes $q_1, \ldots, q_r$ about $\lambda_1, \ldots, \lambda_r$, whose means $\overline{\lambda}_j$ are described by

$$n^{\frac{1}{2}}(\overline{\lambda}_j - \lambda_j) = \frac{1}{q_j} \operatorname{trace} GP_j + O(n^{-\frac{1}{2}}) . \qquad (5.1.12)$$

If the eigenvectors associated with the eigenvalues in the jth cluster are denoted by $\hat{v}_{j1},\ldots,\hat{v}_{jq_j}$ and $\hat{P}_j$ is defined by

$$\hat{P}_j = \hat{v}_{j1}\hat{v}'_{j1}+\ldots+\hat{v}_{jq_j}\hat{v}'_{jq_j} , \tag{5.1.13}$$

then

$$n^{\frac{1}{2}}(\hat{P}_j-P_j) = \sum_{k\neq j} \frac{P_k G P_j + P_j G P_k}{\lambda_j-\lambda_k} + O(n^{-\frac{1}{2}}) . \tag{5.1.14}$$

It will be noted that G occurs linearly in (5.1.12) and (5.1.14), so $\bar{\lambda}_j$ and $\hat{P}_j$ have asymptotically Gaussian distributions. Thus they provide a simple basis for point and confidence region estimation and for testing hypotheses about λ_j and P_j. Rather than pursue this topic in full generality, we will study the density $f(\|x\|_V)$ described below (5.1.8), for which we may get a set of statistical procedures analogous to those given in Chapter 4. Prentice (1982) has considered some of these problems for the case where $\dim \nu = s = 1$. Our account is based on Watson (1982) and references therein.

5.2. Preliminary Results

The density $f(\|x\|_V)$ is that of an antipodally symmetric distribution since the density is equal at x and $-x$. Furthermore, it is invariant under orthogonal transformations within V and within $V^{\perp}$.

We may write

$$x = t\xi + (1-t^2)^{\frac{1}{2}}\eta , \quad t = \|x_V\| , \tag{5.2.1}$$

where $\|\xi\| = 1$, $\xi \in V$, $\|\eta\| = 1$, $\eta \in V^{\perp}$. Along with (5.2.1), we have the decomposition of the measure ω_q on Ω_q,

$$d\omega_q = t^{s-1}(1-t^2)^{\frac{q-s}{2}-1}\, dt\, d\omega_s\, d\omega_{q-s} . \tag{5.2.2}$$

If the density of the law of x is given by $f(\|x_V\|)$, it follows that t, ξ, and

η are independent, that ξ and η are uniformly distributed on unit spheres in $\mathcal{V}$ and $\mathcal{V}^{\perp}$, and that the density of t is given on $(0,1)$ by

$$\omega_s\, \omega_{q-s} f(t)(1-t^2)^{\frac{q-s}{2}-1} t^{s-1} \quad . \tag{5.2.3}$$

Thus f must be non-negative and such that the integral of (5.2.3) from zero to one is unity.

Since $\xi(\eta)$ is uniformly distributed on the unit sphere in $\mathcal{V}(\mathcal{V}^{\perp})$,

$$\begin{aligned} &E(\xi) = 0 \ , \ E(\xi\xi') = P_V/s \ , \\ &E(\eta) = 0 \ , \ E(\eta\eta') = P_{V^{\perp}}/(q-s) \ . \end{aligned} \tag{5.2.4}$$

Thus

$$Ex = 0 \ , \ Exx' = Et^2 \frac{P_V}{s} + (1-Et^2)\frac{P_{V^{\perp}}}{q-s} = M \ , \tag{5.2.5}$$

where Et^2 is to be found by using (5.2.3). The key formulae (5.1.12) and (5.1.14) become

$$n^{\frac{1}{2}}(\bar{\lambda}_1-\lambda_1) = \frac{1}{s} \text{ trace } GP_V + O(n^{-\frac{1}{2}}), \tag{5.2.8}$$

where $\bar{\lambda}_1$ is the mean of the s roots in the cluster about λ_1 and

$$n^{\frac{1}{2}}(\hat{P}_V-P_V) = \frac{P_V G P_{V^{\perp}} + P_{V^{\perp}} G P_V}{\lambda_1-\lambda_2} + O(n^{-\frac{1}{2}}) \ , \tag{5.2.9}$$

where $\lambda_1 = E\|x_V\|^2/s = Et^2/s$, $\lambda_2 = (1-Et^2)(q-s)^{-1}$ are assumed to be unequal. For some calculations, one might need the next terms in the asymptotic expansions (5.2.8) and (5.2.9) which are, respectively, the traces of

$$\frac{-1}{n^{\frac{1}{2}}(\lambda_1-\lambda_2)^2}\left\{\begin{aligned} &\lambda_1 P_V G P_{V_{\perp}} G P_V + \lambda_1 P_{V_{\perp}} G P_V G P_V + \lambda_1 P_V G P_V G P_{V_{\perp}} \\ &+ \lambda_2 P_{V_{\perp}} G P_{V_{\perp}} G P_V + \lambda_2 P_V G P_{V_{\perp}} G P_{V_{\perp}} + \lambda_2 P_{V_{\perp}} G P_V G P_{V_{\perp}} \end{aligned}\right\}$$

and

$$\frac{-1}{n^{\frac{1}{2}}(\lambda_1-\lambda_2)}\left\{\begin{array}{l} P_v GP_{v\perp} GP_v + P_{v\perp} GP_v GP_v + P_v GP_v GP_{v\perp} \\ + P_{v\perp} GP_v GP_{v\perp} + P_v GP_{v\perp} GP_{v\perp} + P_{v\perp} GP_{v\perp} GP_v \end{array}\right\}. \tag{5.2.10}$$

The projector $\hat{P}_v$ is $v_1v_1' + \ldots + v_sv_s'$, where the v_j are the eigenvectors associated with the roots in the cluster about λ_1. While the $\hat{v}_j$ individually become undetermined as $n \longrightarrow \infty$, $\hat{P}_v \longrightarrow P_v$ in probability.

Using (5.2.1) and (5.2.8) and noting that trace GP_v = trace P_vGP_v,

$$\begin{aligned}\sqrt{n}(\bar{\lambda}_1-\lambda_1) &= \frac{1}{s}\, n^{\frac{1}{2}}\left(\frac{1}{n}\sum_1^n t_i^2 - \lambda_1 s\right) + O(n^{-\frac{1}{2}}) \\ &= \frac{1}{s}\, n^{\frac{1}{2}}\left(\frac{1}{n}\sum_1^n t_i^2 - Et^2\right) + O(n^{-\frac{1}{2}}),\end{aligned} \tag{5.2.11}$$

so that by the central limit theorem

$$\mathcal{L}\, n^{\frac{1}{2}}(\bar{\lambda}_1-\lambda_1) \longrightarrow G_1(0,\ s^{-2}\operatorname{var} t^2). \tag{5.2.12}$$

Var t^2 can be computed from (5.2.3), or estimated consistently from the data $x_1,\ldots,x_n$. There is an analogous formula for λ_2. Using (5.2.1) and setting

$$W_n = n^{-\frac{1}{2}}\sum_1^n t_i(1-t_1^2)^{\frac{1}{2}}\xi_i n_i', \tag{5.2.13}$$

we may write (5.2.9) as

$$n^{\frac{1}{2}}(\hat{P}_v-P_v) = \frac{1}{\lambda_1-\lambda_2}(W_n+W_n') + O(n^{-\frac{1}{2}}). \tag{5.2.14}$$

Note that since $P_v + P_{v\perp} = I_q$,

$$n^{\frac{1}{2}}(\hat{P}_{v\perp}-P_{v\perp}) = -n^{\frac{1}{2}}(\hat{P}_v-P_v). \tag{5.2.14}$$

Then

$$n\|\hat{P}_v-P_v\|^2 = \frac{2}{(\lambda_1-\lambda_2)^2}\operatorname{trace} W_nW_n' + O(n^{-\frac{1}{2}}) \tag{5.2.16}$$

because $W_n W_n = 0$.

By the MCLT, as $n \longrightarrow \infty$, $W_n \longrightarrow W$ where the elements of W are jointly Gaussian. We may write the elements of W as a column vector C_n partitioned into the first, second, ..., qth columns of W, in that order.

Then $C_n \longrightarrow G_{q^2}(0,\Sigma_c)$. The matrix Σ_c can be found from (5.2.4), and may also be thought of as the dispersion matrix of W. To find Σ_c, we recognize that it is the dispersion matrix of $t(1-t^2)^{\frac{1}{2}}\ \xi \otimes \eta$, where $\otimes$ is the left direct matrix product.

Hence

$$\Sigma_c = \frac{(Et^2-Et^4)}{s(q-s)}\ P_v \otimes P_{v_\perp}. \tag{5.2.17}$$

Observe that $P_v \otimes P_{v_\perp}$ is an idempotent with $s(q-s)$ unit eigenvalues. Returning to (5.2.15), we see that

$$n\|\hat{P}_v-P_v\|^2 \longrightarrow \frac{2(Et^2-Et^4)}{(\lambda_1-\lambda_2)^2}\ \frac{\chi^2_{s(q-s)}}{s(q-s)} \quad . \tag{5.2.18}$$

Suppose that H is an orthogonal matrix which makes the first s coordinates span V and the last $q-s$ span $V_\perp$ Then (5.2.13) shows that HW_nH' is all zeros except for the top right $s\times(q-s)$ submatrix U_n', whose entries have zero means, are uncorrelated, and have variances $(Et^2-Et^4)/s(q-s)$ by (5.2.17). Then the eigenvalues of (W_n+W_n'), satisfy

$$\begin{vmatrix} -\lambda I_s & U_n' \\ U_n & -\lambda I_{q-s} \end{vmatrix} = 0 \quad . \tag{5.2.19}$$

This determinant may also be written as

$$|-\lambda I_s|\,|U_nU_n' - \lambda I_{q-s}| = |-\lambda I_{q-s}|\,|U_n'U_n - \lambda I_s|. \tag{5.2.20}$$

We will suppose that $s \leq q - s$. Thus $W_n + W_n'$ has $q - s$ zero eigenvalues, and the remaining s are the roots of $|U_n'U_n - \lambda I_s|$. As $n \longrightarrow \infty$, $U_n \longrightarrow U$ where the matrix $U(q-s) \times s)$ has independent $G_1(0, \frac{Et^2 - Et^4}{s(q-s)})$ entries and $U'U$ has a Wishart distribution.

But the distribution of the eigenvalues of a Wishart matrix is known, so that by (5.2.14), we know the asymptotic distribution of the eigenvalues of $n^{\frac{1}{2}}(P_V - \hat{P}_V)$. Since (5.2.14) may now be written

$$n^{\frac{1}{2}}H(\hat{P}_V - P_V)H' = \frac{1}{(\lambda_1 - \lambda_2)} \begin{bmatrix} 0 & U' \\ U & 0 \end{bmatrix} + O(n^{-\frac{1}{2}}) , \tag{5.2.21}$$

$$n\|\hat{P}_V - P_V\|^2 = \frac{2}{(\lambda_1 - \lambda_2)} \text{ trace } U'U, \tag{5.2.22}$$

we have another proof of (5.2.18). It may be noted that U' is equal to the uppe right $s \times (q-s)$ submatrix of $HP_V GP_{V^\perp}H')$.

The s eigenvalues just discussed will be given a geometrical interpretatic in Section 5.3. They have no direct relationship to the s eigenvalues of M_n that cluster about λ_1. Denoting these latter roots by $\hat{\lambda}_1, \ldots, \hat{\lambda}_s$, they satisfy $M_n\hat{v}_j = \hat{\lambda}_j\hat{v}_j$ $(i=j,\ldots,s)$. The s non-zero eigenvalue of $n^{\frac{1}{2}}(\hat{P}_V M_n \hat{P}_V - \lambda_1 \hat{P}_V)$ are $n^{\frac{1}{2}}(\hat{\lambda}_j - \lambda_1)$, $i = 1,\ldots,s$. Using (5.2.7) and (5.2.9) $n^{\frac{1}{2}}(\hat{P}_V M_n \hat{P}_V - \lambda_1 \hat{P}_V)$ is asymptotically equal to $P_V GP_V = F$, say a matrix with Gaussian elements with zero means and variances and covariances which are those of $t^2\xi\xi'$. If we consider instead HFH', we need only consider the top left $s \times s$ submatrix or regard ξ as an s-vector uniformly distributed on the unit sphere Ω_s, so that $E\xi = 0$, $E\xi\xi' = I_s/s$. Writing the (k,ℓ) element of $t^2\xi\xi'$ as $a_{k\ell} = t^2\xi_k\xi_\ell$,

$$Ea_{kk} = Et^2/s \;, \; Ea_{k\ell} = 0 \quad (k \neq \ell), \tag{5.2.23}$$

$$Ea_{kk}^2 = Et^4E\xi_k^4 = Et^4\frac{3}{s(s+2)} \;,$$

$$Ea_{k\ell}^2 = Et^4E\xi_k^2\xi_\ell^2 = Et^4\frac{1}{s(s+2)} = Ea_{kk}a_{\ell\ell}, \tag{5.2.24}$$

$$Ea_{ij}a_{k\ell} = 0 \;, \text{ otherwise,}$$

by using results in Anderson & Stephens (1972). Hence we may write down the covariance matrix of the $s(s+1)/2$ functionally independent elements $a_{11},\dots,a_{ss}, a_{12},\dots,a_{1s},\dots,a_{s-1s}$. Then

$$\begin{aligned} n\sum_1^s (\hat{\lambda}_j - \lambda_1)^2 &\sim \|P_V G P_V\|^2 = \|HFH'\|^2 \\ &= \sum_1^s (HFH')^2_{kk} + 2\sum_{k<l}^s (HFH')^2_{k\ell} \,, \end{aligned} \tag{5.2.25}$$

a sum of squares of Gaussians whose covariance matrix has just been determined. However (5.2.25) does not have the simple distribution here that it does when $Lx=U(\Omega_q)$, the case considered in Anderson & Stephens.

In practice however, we will need the asymptotic distribution of $n\sum_1^s (\hat{\lambda}_j - \bar{\lambda}_1)^2$, where $\bar{\lambda}_1 = s^{-1}(\hat{\lambda}_1+\dots+\hat{\lambda}_s)$. The $n^{\frac{1}{2}}(\hat{\lambda}_j - \bar{\lambda}_1)$ are non-zero eigenvalues of

$$K = P_V G P_V - \frac{(\text{trace } GP_V)}{s} P_V, \tag{5.2.26}$$

upon using (5.2.8). Using the previous paragraph, we are led to a symmetric matrix HKH' with Gaussian entries with means and variances and covariances equal to those of

$$[b_{k\ell}] = t^2(\xi\xi' - I_s/s) = t^2[\xi_k\xi_\ell - \delta_{k\ell}/s]. \tag{5.2.27}$$

Clearly, $Eb_{k\ell} = 0$ and

$$\left.\begin{aligned} \operatorname{var} b_{kk} &= Et^4\left(\frac{3}{s(s+2)} - \frac{1}{s^2}\right) = Et^4\,\frac{2(s-1)}{s^2(s+2)}, \\ \operatorname{var} b_{k\ell} &= Et^4\,\frac{1}{s(s+2)}\,, \\ \operatorname{cov}(b_{kk},b_{k\ell}) &= Et^4\,\frac{-2}{s^2(s+2)},\ (k \neq \ell)\,, \\ \operatorname{cov}(b_{ij},b_{k\ell}) &= 0\,,\ \text{otherwise}\,. \end{aligned}\right\} \tag{5.2.28}$$

Since

$$n(\hat{\lambda}_j - \bar{\lambda}_1)^2 = \sum_1^s (HKH')^2_{kk} + 2 \sum_{k<\ell} (HKH')^2_{k\ell} ,$$

the argument used by Anderson & Stephens (1972) shows that

$$\mathcal{L}\, n \sum_1^s (\hat{\lambda}_j - \bar{\lambda}_1)^2 \longrightarrow \frac{2Et^4}{s(s+2)} \chi^2_{\frac{s(s+1)}{2} - 1} . \qquad (5.2.29)$$

5.3. Inferences from One Sample

The primary problems are to estimate V, to provide a confidence region, and to test that $V = V_0$. The latter procedures require consistent estimators of Et^2 and Et^4, and these are provided by $n^{-1}\Sigma\|\hat{P}_v x_i\|^2$ and $n^{-1}\Sigma\|\hat{P}_v x_i\|^4$. But one often wishes to check the rotational symmetry implied by $f(\|x_v\|)$ and so to check the equality of the cluster of eigenvalues associated with λ_1. $V(\hat{V})$ is uniquely identified by $P_v(\hat{P}_v)$. The estimator $\hat{P}_v$ (defined below (5.2.10)) is seen by (5.2.9) to be consistent. A mathematically natural confidence region is given by using (5.2.18) but its practical interpretation is difficult, and we will later seek an alternative; (5.2.18) also provides a test of the null hypothesis $V = V_0$. An apparently different test statistic is, where $P_{v0}V_0 = V_0$,

$$\begin{aligned} T &= n(\max_P \text{trace } PM_n - \text{trace } P_{v0}M_n) \\ &= n \text{ trace } (\hat{P}_v - P_{v0})M_n . \end{aligned} \qquad (5.3.1)$$

But using (5.2.9) with V_0 instead of V, we see that

$$T = \frac{2}{\lambda_1 - \lambda_2} \text{ trace } (P_{v0}GP_{v0})(P_{v0}GP_{v0})' + O(n^{-\frac{1}{2}}), \qquad (5.3.2)$$

$$n\|\hat{P}_v - P_{v0}\|^2 = T + O(n^{-\frac{1}{2}}) \qquad (5.3.3)$$

so the two tests are equivalent as $n \longrightarrow \infty$. We will need this result in Section 5

An important special case is $s = 1$ when we will write $P_V = \mu\mu'$, $\hat{P}_V = \hat{\mu}\hat{\mu}'$ when μ $(\hat{\mu})$ is the unit eigenvector associated with the largest eigenvalue of $M(\hat{M})$, (5.2.18) then reads

$$2n(1-\hat{\mu}'\mu) \sim \frac{2(Et^2 - Et^4)}{(\lambda_1 - \lambda_2)^2} \frac{\chi^2_{q-1}}{q-1} . \tag{5.3.4}$$

Writing $\hat{\mu}'\mu = \cos\hat{\theta}$, (5.3.4) provides a "circular" confidence cone for μ with semi-angle $\hat{\theta}$, when estimators are inserted, as well as a test that the true axis is μ. To obtain the power of this test, as $n \longrightarrow \infty$, we set

$$\mu = \mu_0 + \delta n^{-\frac{1}{2}} , \delta \perp \mu_0 . \tag{5.3.5}$$

From (5.2.13) with $s = 1$,

$$n^{\frac{1}{2}}(\hat{\mu} - \mu) = \frac{n^{-\frac{1}{2}}}{\lambda_1 - \lambda_2} \sum_1^n t_i(1-t_i^2)^{\frac{1}{2}}\eta_i + O(n^{-\frac{1}{2}}),$$

so that

$$n^{\frac{1}{2}}(\hat{\mu} - \mu_0) = -\delta + \frac{n^{-\frac{1}{2}}}{\lambda_1 - \lambda_2} \sum_1^n t_i(1-t_i^2)^{\frac{1}{2}}\eta_i + O(n^{-\frac{1}{2}}).$$

Thus as $n \longrightarrow \infty$,

$$\mathcal{L}\, n^{\frac{1}{2}}(\hat{\mu} - \mu_0) \longrightarrow G_q(-\delta, \frac{Et^2 - Et^4}{(\lambda_1-\lambda_2)^2} \frac{I_q - \mu_0\mu_0'}{q-1}) \tag{5.3.6}$$

and

$$\mathcal{L}\, 2n(1-\hat{\mu}'\mu_0) \longrightarrow \frac{2(Et^2-Et^4)}{(\lambda_1-\lambda_2)^2(q-1)} \chi^2_{q-1}\{\frac{\|\delta\|^2(q-1)(\lambda_1-\lambda_2)^2}{Et^2-Et^4}\}. \tag{5.3.7}$$

The power of the test $V = V_0$ for general s may be derived similarly by setting

$$P_V = P_{V_0} + Q\, n^{-\frac{1}{2}},$$
$$Q = Q' , Q = P_{V_0}Q + QP_{V_0} . \tag{5.3.8}$$

The conditions ensure that P_v is symmetric and idempotent to $0(n^{-\frac{1}{2}})$. We find that

$$\mathcal{L}\, n\|\hat{P}_v - P_{v_0}\|^2 \longrightarrow \frac{2(Et^2-Et^4)}{(\lambda_1-\lambda_2)^2 s(q-s)} \chi^2_{s(q-s)}\left\{\frac{\|Q\|^2 s(q-1)(\lambda_1-\lambda_2)}{Et^2-Et^4}\right\} . \qquad (5.3.9)$$

Another way of analysing the difference between V and $\hat{V}$ is to consider the angle between vectors $v = P_v u$ and $\hat{v} = \hat{P}_v \hat{u}$ in V and $\hat{V}$, respectively. The stationary values of the cosine of this angle may be shown to satisfy the determinantal equation

$$\begin{vmatrix} -cP_v & P_v\hat{P}_v \\ \hat{P}_v P_v & -cP_v \end{vmatrix} = 0 \quad ; \qquad (5.3.10)$$

we will suppose that $s \le q - s$.

Using a g-inverse of P_v, (5.3.10) may be reduced to

$$|P_v\hat{P}_v P_v - c^2 P_v| = 0, \qquad (5.3.11)$$

which has s non-zero roots c_j^2. The expansion (5.2.9) and (5.2.10) of P_v can be written

$$\hat{P}_v = P_v + n^{-\frac{1}{2}}\Delta_1 + n^{-1}\Delta_2 + 0(n^{-\frac{3}{2}}). \qquad (5.3.12)$$

Since

$$P_v\Delta_1 P_v = 0 \ ,$$

$$P_v\Delta_2 P_v = \frac{-P_v G P_{v\perp} G P_v}{(\lambda_1 - \lambda_2)^2} \ . \qquad (5.3.13)$$

We deduce from (5.3.12), since P_v and $P_v \hat{P}_v P_v$ can be simultaneously diagonalized, that

$$\mathcal{L}_{n(1-c_j)} \longrightarrow \mathcal{L}_{2n(1-c_j)}$$
$$= \mathcal{L}_{\frac{1}{(\lambda_1 - \lambda_2)} \lambda_j (WW')},$$

(5.3.14)

where W appeared before below (5.2.16), (5.2.18) gives the distribution of the sum of the $\lambda_j(WW')$, or trace WW'.

Finally, to test the rotational symmetry within V, the natural statistic is $n^{\frac{1}{2}} \sum_1^s (\hat{\lambda}_j - \bar{\lambda}_1)^2$, whose asymptotic distribution is given at the end of Section 2 - specifically by (5.2.29). To use this result in practice, one may replace Et^4 by the consistent estimator $n^{-1} \Sigma \|\hat{P}_v x_i\|^4$.

4. Inferences from several samples

We now suppose large samples of $n_1, n_2, \ldots, n_m$ are available from m populations with densities depending only upon $\|x_{v1}\|, \ldots, \|x_{vm}\|$ where $V_1, \ldots, V_m$ are s-dimensional subspaces. We will write $n_i = \alpha_i N$, $\alpha_i > 0$, $\Sigma \alpha_i = 1$ and let $N \longrightarrow \infty$. The densities will be written $f_1(\|x_{v1}\|), \ldots, f_m(\|x_{vm}\|)$ but Et^2, Et^4, λ_1, λ_2 will be assumed to be the same in each population and we will suppose that $\lambda_1 > \lambda_2$. The data from each sample will be summarized by M_{nj}, $j=1, \ldots, m$.

For distributions with densities of the form $f(\|x_v\|)$, the concentration could be measured by $Et^2 = s\lambda_1$. Thus comparisons of concentrations are comparisons of λ_1's, and the result (5.2.12) could be used.

To construct a test of the null hypothesis H that $V_1 = V_2 = \ldots = V_m$, a plausible statistic is

$$\frac{T_N}{N} = \sum_1^m \alpha_j \max_{P_{vj}} \text{trace } P_{vj} M_{nj} - \max_{P_v} \text{trace } P_v \sum_1^m \alpha_j M_{nj} . \tag{5.4.1}$$

Clearly, $T_N \geq 0$ should be large when the true P_{vj} are distinct, and may be rewritten as

$$\frac{T_N}{N} = \sum_{j=1}^m \alpha_j \{\lambda_1(M_{nj})+\ldots+\lambda_s(M_{nj})\} - \{\lambda_1(\Sigma\alpha_j M_{nj})+\ldots+\lambda_s(\Sigma\alpha_j M_{nj})\} , \tag{5.4.2}$$

where $\lambda_1 > \lambda_2 > \ldots > \lambda_q$; (5.4.1) generalizes (5.3.1) to many samples, and (5.4.2) is its computational form. We first consider the distribution of (5.4.1) when the null hypothesis $V_1 = \ldots = V_m$ is true.

Since $\Sigma\alpha_j = 1$,

$$\Sigma\alpha_j M_{nj} = M + N^{-\frac{1}{2}} G^* , \tag{5.4.3}$$

where

$$M_{nj} = M + N^{-\frac{1}{2}} G_j \alpha_j^{-\frac{1}{2}} , \quad G^* = \Sigma\alpha_j^{\frac{1}{2}} G_j . \tag{5.4.4}$$

Suppose

$$\max_{P_v} P_v \Sigma\alpha_j M_{nj} = P^* \Sigma\alpha_j M_{nj} \tag{5.4.5}$$

so that

$$\frac{T_N}{N} = \sum_1^m \alpha_j \quad \text{trace } (\hat{P}_{vj} - P^*) M_{nj} . \tag{5.4.6}$$

Using (5.2.9) twice,

$$\hat{P}_{vj} - P^* = N^{-\frac{1}{2}}(\lambda_1-\lambda_2)^{-1}\{P_v(G_j\alpha_j^{-\frac{1}{2}}-G^*)P_{v\perp} + P_{v\perp}(G_j\alpha_j^{-\frac{1}{2}}-G^*)P_v\} + O(N^{-1}). \tag{5.4.7}$$

Hence (5.4.4) and (5.4.7) in (5.4.6) yield the results:

$$\mathcal{L}T_N \longrightarrow \mathcal{L}\sum_1^m \frac{\alpha_j}{\lambda_1-\lambda_2} \text{ trace } \{P_v(\frac{G_j}{\alpha_j^{\frac{1}{2}}} - G^*)P_{v\perp}\frac{G_j}{\alpha_j^{\frac{1}{2}}} + P_{v\perp}(\frac{G_j}{\alpha_j^{\frac{1}{2}}} - G^*)P_v\frac{G_j}{\alpha_j^{\frac{1}{2}}}, \tag{5.4.8}$$

or

$$\mathcal{L}T_N\frac{(\lambda_1-\lambda_2)}{2} \longrightarrow \mathcal{L}\{\sum_1^m \text{ trace } P_vG_jP_{v\perp}G_j - \text{trace } P_vG^*P_v\, G^*\}. \tag{5.4.9}$$

Setting

$$B_j = HP_vG_jP_v\, H', \tag{5.4.10}$$

where the orthogonal matrix H makes $P_v(P_{v_\perp})$ a diagonal matrix beginning (ending) with $s(q-s)$ unities, the random variable on the right-hand side of (5.4.10) may be rewritten as

$$\text{trace } B_jB_j' - \text{trace } (\Sigma\alpha_j^{\frac{1}{2}}B_j)(\Sigma\alpha_j^{\frac{1}{2}}B_j)'. \tag{5.4.11}$$

But we saw, in the derivation of (5.2.22), that the non-zero elements of B_j could be called U_j' and that all the elements of $U_1,\ldots,U_m$ are, independently $G_1(0,\psi)$,

$$\psi = (Et^2-Et^4)s^{-1}(q-s)^{-1} > 0. \tag{5.4.12}$$

Further (5.4.11) is true if B_j is replaced by U_j, and one may now guess the choice $\omega_j = 1$ will lead to (5.4.11) being a multiple of a Chi-square. Defining matrices with independent $G_1(0,1)$ entries,

$$Y_j = \psi^{-\frac{1}{2}} U_j, \tag{5.4.13}$$

we need the distribution of

$$\psi\{\sum_1^m \text{trace } Y_j Y_j' - \text{trace } (\Sigma\alpha_j^{\frac{1}{2}} Y_j)(\Sigma\alpha_j^{\frac{1}{2}} Y_j)'\}. \tag{5.4.14}$$

If $\{K_{ij}\}$ is an $m \times m$ orthogonal array with $K_{mj} = \alpha_j^{\frac{1}{2}}$ (possible since $\Sigma\alpha_i = 1$), and

$$Z_i = K_{i1}Y_1 + \dots + K_{im}Y_m \ , \ i = 1,\dots,m \ , \tag{5.4.15}$$

then the elements of $Z_1,\dots,Z_m$ are independent $G_1(0,1)$, and (5.4.14) may be rewritten as

$$\psi \sum_1^{m-1} \text{trace } Z_i Z_i' = \psi\chi^2_{(m-1)s(q-s)}. \tag{5.4.16}$$

Collecting results, the asymptotic distribution of

$$\begin{aligned} &\sum_1^m \max_{P_{vj}} P_{vj} M_{nj} - \max_{P_v} P_v \sum_1^m n_j M_{nj} \\ &= \sum_n^m n_j\{\lambda_1(M_{nj}) + \dots + \lambda_s(M_{nj})\} \\ &- \{\lambda_1(\Sigma_{nj} M_{nj}) + \dots + \lambda_s(\Sigma_{nj} M_{nj})\} \end{aligned} \tag{5.4.17}$$

is, when the null hypothesis $V_1 = \dots = V_m$ is true, that of

$$\frac{2(Et^2 - Et^4)}{qEt^2 - s} \chi^2_{(m-1)s(q-s)}. \tag{5.4.18}$$

To use this test in practice, the consistent estimators $n^{-1}\Sigma\|\hat{P}_v x_i\|^2$ and $n^{-1}\Sigma\|\hat{P}_v x_i\|^4$ will be needed. The second form in (5.4.18) is to be used computationally. It is shown in Watson (1982) that on the alternative hypothesis,

$$M_{nj} = M_j + N^{-\frac{1}{2}} G_j \alpha_j^{-\frac{1}{2}} , \; M_j = \lambda_1 P_{vj} + \lambda_2 P_{vj_\perp} \tag{5.4.19}$$

for $j = 1, \ldots, m$, that the limiting distribution of (5.4.1) is

$$\frac{2(Et^2 - Et^4)}{\lambda_1 - \lambda_2} \chi^2_{(m-1)s(q-s)}(\theta), \tag{5.4.20}$$

where the non-centrality parameter θ is given by

$$\theta = \psi^{-1}(\lambda_1 - \lambda_2)^2 \{\Sigma_j \text{ trace } Q_j Q_j' - \text{trace}(\Sigma \alpha_j^{\frac{1}{2}} Q_j)(\Sigma \alpha_j^{\frac{1}{2}} Q_j)'. \tag{5.4.21}$$

To get his limit we suppose that

$$\begin{aligned} P_{vj} &= P_v + N^{-\frac{1}{2}} Q_j \alpha_j^{-\frac{1}{2}} , \\ Q_j &= Q_j' , \; Q_j = P_v Q_j + Q_j P_v \; . \end{aligned} \tag{5.4.22}$$

5.5 <u>The Generalized Scheiddegger-Watson Distribution</u>

A special case of $f(\|x_v\|)$ is the density

$$\frac{\exp \kappa \|x_v\|^2}{\omega_q \, M(\frac{s}{2}, \frac{q}{2}, \kappa)} \quad (\kappa \geq 0) , \tag{5.5.1}$$

where $M(a,b,z)$ is the confluent hypergeometric function, known after Kummer, and defined in Section 13.1 of Abramowitz and Stegun (1965), and where $\omega_q = 2\pi^{q/2}/\Gamma(q/2)$ is the area of Ω_q.

Given a sample of n from (5.5.1), it is clear that the m.ℓ. estimates $\hat{P}_v$ and $\hat{\kappa}$ are defined by

$$\hat{P}_v = \hat{v}_1\hat{v}_1' + \ldots + \hat{v}_s\hat{v}_s' , \qquad (5.5.2)$$

$$\frac{\partial}{\partial\hat{\kappa}} \log M(\tfrac{s}{2}, \tfrac{q}{2}, \hat{\kappa}) = \lambda_1(M_n) + \ldots + \lambda_s(M_n) , \qquad (5.5.3)$$

where $\lambda_1(M_n),\ldots,\lambda_s(Mn)$ are the s largest eigenvalues of Mn, and $\hat{v}_1,\ldots,\hat{v}_s$ are their associated eigenvectors. It may be shown that (5.5.3) always has a unique solution, and that it may be rewritten as

$$S(\hat{k}) = \frac{M(\frac{s}{2}+1,\frac{q}{2}+1,\hat{\kappa})}{q\,M(\frac{s}{2},\frac{q}{2},\hat{\kappa})} = \bar{\lambda}_1 , \qquad (5.5.3')$$

where $\bar{\lambda}_1$ is the arithmetic mean of $\lambda_1(M_n),\ldots,\lambda_s(M_n)$. It follows from (5.2.8), (5.2.12) that

$$\mathcal{L}\, n^{\frac{1}{2}}(\hat{\kappa}-\kappa)S'(\kappa) \longrightarrow \mathcal{L}\, \frac{1}{s} \text{ trace } GP_v, \qquad (5.5.4)$$

$$= G_1(0, b^{-2} \text{ var } t^2) . \qquad (5.5.5)$$

This result could be used in practice if $S'(\kappa)$ and var t. are evaluated using $\hat{\kappa}$ instead of κ. Also a variance stabilizing transformation for $\hat{\kappa}$ could be derived from (5.5.5).

For samples from m distributions (5.5.1) with the same κ, a test of $V_1 = \ldots = V_m$ could be derived by specializing the results of the last section. The test statistic used there is, in fact, the likelihood ratio test for data from the distribution (5.5.1) when κ is assumed known. It is asymptotically equivalent to the likelihood ratio test when κ is not known because $\hat{\kappa}$ and $\hat{P}_v$ are asymptotically independent, a fact which may be verified by using (5.2.8) and (5.2.9).

By analogy with Chapter 4, it would be interesting to see here if the methods become simpler as the concentrations in the distributions become larger, but this has not yet been done.

5.6 Application to the Langevin Distribution

When the density of x has the Langevin form

$$a_q(\kappa)^{-1} \exp \kappa\mu' x \,, \tag{5.6.1}$$

it was shown in Chapter 4 that the m.l. estimators of μ and κ are respectively $\hat{\mu} = X/\|X\|$ and $\hat{\kappa}$, the solution of

$$A_q(\hat{\kappa}) = \|x\| / n \,,$$

or

$$\coth \hat{\kappa} - \hat{\kappa}^{-1} = \|X\| / n \,, \quad \text{if } q = 3$$

However when (5.6.1) holds,

$$M = Exx' = (A_q' + A_q^2)\mu\mu' + \frac{A_q}{\kappa}(I - \mu\mu'), \tag{5.6.2}$$

where of course

$$A_q(\kappa) = \frac{d}{d\kappa} \log a_q(\kappa) \,.$$

Thus the methods of this chapter can be used to find the relative efficiencies of estimators of μ and κ, μ^* and κ^*, say, derived from M_n. These are respectively the eigenvector of M_n, $\hat{v}_1$ associated with its isolated largest eigenvalue $\hat{\lambda}_1$. While $\mu^* = \hat{v}_1$, to find κ^* we must solve

$$A_q'(\kappa^*) + A_q^2(\kappa^*) = \hat{\lambda}_1 \,. \tag{5.6.3}$$

To carry out the computations when $n \to \infty$, we may use (5.2.12) and the fact that $\kappa^* \to \kappa$ in probability to find the variance of κ^* as a function of κ. The variance, as $n \to \infty$, of $\hat{\kappa}$ is given in Chapter 4. The relative efficiency of κ^* is then $\mathrm{var}(\hat{\kappa})/\mathrm{var}(\kappa^*)$.

Similarly, the asymptotic distributions of μ^* and $\hat{\mu}$ may be found from the results of Chapters 4 and 5. The projections of μ^* and $\hat{\mu}$ onto the tangent space of Ω_q at μ will be spherically symmetric Gaussian distributions with mean vectors zero and covariance matrices $\alpha^*(I-\mu\mu')$ and $\hat{\alpha}(I-\mu\mu')$, say. A natural definition of the relative efficiency of μ^* is then the ratio $\hat{\alpha}/\alpha^*$.

Javier Cabrera made these studies at the suggestion of M.J. Prentice and G.S. Watson. The details will appear elsewhere.

The relative efficiency of κ^* was computed for $q=3,4,5$ and κ in the range $(0,16)$. It is zero when $\kappa=0$. For $\kappa > 10$, the 3 curves are almost indistinguishable and greater than .97. At $\kappa = 4$, the efficiencies are, respectively .84, .82, .80. We saw in Chapter 1 that for $q=3$, $\kappa=4$ was a relatively small value and that it corresponds to a very dispersed distribution. In palaeomagnetic examples, κ's much larger than 10 are the rule.

The relative efficiency of μ^* has so far only been computed for $q=3$. The curve rises from zero at $\kappa=0$, to 0.81 at $\kappa=4$, and to .97 at $\kappa=10$.

These results suggest that it may be wiser to use the estimators μ^* and κ^* since there are reasons for thinking that they are more robust. For example, in the palaeomagnetic case, reversals $x \to -x$ are a feature of some data sets. These leave μ^* and κ^* unchanged. Thus despite the long history in palaeomagnetic statistics of the use of $X/\|X\|$ and $(n-1)/(n - \|X\|)$ as estimator of μ and κ, it seems that μ^* and κ^* should be used!

Appendix A Normalization Constants

1. INTRODUCTION

Two probability densities on the surface Ω_q of the unit ball in $\mathbb{R}^q$ mentioned frequently in the text above are:

$$f_1(x) = a_q^*(\kappa)^{-1} \exp \kappa\, \mu' x , \tag{1.1}$$

$$f_2(x) = b_q^*(\kappa)^{-1} \exp \kappa (\mu' x)^2 \tag{1.2}$$

when $\mu' x$ is the scalar product of the two unit vectors x and μ . μ is called the modal direction and κ a concentration parameter. In (1.1), $\kappa \geqslant 0$. In (1.2), κ may be any real number. If $\kappa \to 0$, either density becomes uniform.

The statistical theory of (1.1), the Langevin distribution, has been discussed in Chapter 4. The density of (1.2), the Scheidegger-Watson distribution, is a special case of that discussed in Chapter 5. It was found that the theory turns upon

$$A_q(\kappa) = a_q^{*\prime}(\kappa)/a_q^*(\kappa) , \tag{1.3}$$

$$B_q(\kappa) = b_q^{*\prime}(\kappa)/b_q^*(\kappa) , \tag{1.4}$$

respectively. The integral definitions of a_q^* and b_q^* were simplified by writing $x = t\mu + (1-t^2)^{\frac{1}{2}}\xi$, where $t = \mu' x$ and ξ is a unit vector orthogonal to μ . Then, writing

$$\omega_q = \text{area of } \Omega_q = 2\pi^{q/2}/\Gamma(q/2) , \tag{1.5}$$

we have

$$a_q^*(\kappa) = \omega_{q-1} a_q(\kappa) , \; b_q^*(\kappa) = \omega_{q-1}\, b_q(\kappa) ,$$

where

$$a_q(\kappa) = \int_{-1}^{1} e^{\kappa t}(1-t^2)^{(q-3)/2} dt \quad , \tag{1.6}$$

$$b_q(\kappa) = \int_{-1}^{1} e^{\kappa t^2}(1-t^2)^{(q-3)/2} dt \quad , \tag{1.7}$$

and

$$A_q(\kappa) = a_q'(\kappa)/a_q(\kappa) \ , \ B_q(\kappa) = b_q'(\kappa)/b_q(\kappa) . \tag{1.8}$$

In Physics, $A_q(\kappa)^1$ is referred to as the Langevin function. Dyson, Lieb and Simon (1978), in proving the existence of spontaneous magnetization at sufficiently low temperatures, needed some properties $\log a_q(\kappa)$. Suppose $h(y)$ is defined by

$$h(y) = \log \int_{\Omega_q} \exp y'x \, d\mu(x),$$

where $\mu(x)$ is any measure on Ω_q . Then $h(y)$ is a convex function of y . If c is the largest eigenvalue of $[\partial^2 h/\partial y_i \partial y]$, $-\frac{1}{2} cy'y + h(y)$ is concave in y . If μ is the uniform measure on Ω_q , $c=q^{-1}$, a proof of which is given below. Their first result is basic for Laplace transforms -- see, e.g., Barndorff-Nielsen (1978).

Schou (1978) gave some properties of and expansions for $A_q(\kappa)$ by recognizing that $a_q(\kappa)$ is proportional $I_{(q/2)-1}$, $(\kappa)\kappa^{-(q/2)-1}$ and using the properties of the modified Bessel function of the first kind, $I_\nu(\kappa)$. Watson (1956) gave some results on $B_q(\kappa)$, q=3 .

We will derive all the properties of $A_q(\kappa)$ (Section 2) and $B_q(\kappa)$ (Section 3) directly, rather than as the ratios a_q'/a, b_q'/b. More properties and more terms in the expansions will be given than heretofore. Furthermore in practice one needs to be able to compute any q and κ, $A_q(\kappa)$, $B_q(\kappa)$, solve the equations $y = A_q(\kappa)$, $y = B_q(\kappa)$ and to apply the variance stabilizing transformations $g_q(\kappa) = \int^{\kappa} A_a'(\kappa)^{\frac{1}{2}} dk$, $h_q(\kappa) = \int^{\kappa} B_q'(\kappa)^{\frac{1}{2}} dk$ numberically. These matters are discussed in Section 4.

While we have wished to emphasize that it is A_q and B_q rather than a_q and b_q which matter, we have for completeness included all known properties of the functions a_q and b_q. The former is associated with the modified Bessel function of the first kind $I_\nu(z)$ and the latter to the less well known Kummer function of the first kind $M(a,b,z)$.

2. The functions $a_q(\kappa)$, $A_q(\kappa)$ and its inverse.

Consider the functions of $\kappa \geqslant 0$.

$$a_q(\kappa) = \int_{-1}^{1} e^{\kappa t}(1-t^2)^{(q-3)/2}dt , \quad q \geqslant 2 , \tag{2.1}$$

$$A_q(\kappa) = a_q'(\kappa)/a_q(\kappa) . \tag{2.2}$$

For the applications now envisaged, $q \geqslant 2$ is an integer, but the following results are true for real $q \geqslant 2$.

It is clear that if $q_1 > q_2$, $0 < a_{q_1}(\kappa) < a_{q_2}(\kappa)$. It is simpler to work with

$$y_\nu(\kappa) = a_q(\kappa) , \quad \nu = (q-3)/2 \geqslant -\tfrac{1}{2} , \tag{2.3}$$

$$M_\nu(\kappa) = y_\nu'(\kappa)/y_\nu(\kappa) = A_q(\kappa) . \tag{2.4}$$

Examination of

$$y_\nu'(\kappa) = \int_{-1}^{1} te^{\kappa t}(1-t^2)^\nu dt$$

shows that

$$0 \leqslant y_\nu'(\kappa) \leqslant y_\nu(\kappa) , \tag{2.5}$$

whence

$$0 \leqslant A_q(\kappa) = M_\nu(\kappa) \leqslant 1 . \tag{2.6}$$

Elementary manipulations show that

$$y_\nu' = \frac{\kappa}{2(\nu+1)} y_{\nu+1} , \quad (\nu > -1) , \tag{2.7}$$

$$\kappa y_\nu' = -(2\nu+1)y_\nu + 2\nu\, y_{\nu-1} \;,\; (\nu > 0) \;, \tag{2.8}$$

$$\kappa y_\nu'' + 2(\nu+1)y_\nu' - \kappa y_\nu = 0 \;,\; (\nu \geq -\tfrac{1}{2}) \;. \tag{2.9}$$

It may be verified that (2.9) has a solution proportional to $I_{\nu+\frac{1}{2}}(\kappa)\,\kappa^{-(\nu+\frac{1}{2})}$ whence

$$a_q(\kappa) = (2\pi)^{q/2}\, I_{q/2-1}(\kappa)\, \kappa^{-q/2+1} \;. \tag{2.10}$$

Here $I_\nu(\kappa)$ is the modified Bessel function of the first kind as defined in Watson (1952). Thus, as $\kappa \to 0$,

$$a_q(\kappa) = (2\pi)^{q/2}\, \kappa^{q/2-1} \sum_{r=0}^{\infty} \frac{(\kappa/2)^{q/2-1+2r}}{r!(q/2-1+r)!} \;, \tag{2.11}$$

while, as $\kappa \to \infty$,

$$a_q(\kappa) = (2\pi)^{q/2-1}\, \kappa^{(q-3)/2}\, e^{\kappa}(1 + O(\kappa^{-1})) \;. \tag{2.12}$$

Since

$$A_q'(\kappa) = a_q''/a_q - (a_q'/a_q)^2 \;, \tag{2.13}$$

$$a_q'' < a_q \;,$$

it follows that

$$A_q'(\kappa) + A_q^2(\kappa) < 1 \;. \tag{2.14}$$

Using the Cauchy inequality, $a_q''a_q \geq (a_q')^2$, in (2.13) and the result of Dyson, Lieb and Simon (1978), we have

$$0 \leq A_q'(\kappa) \leq 1/q \;. \tag{2.15}$$

Returning the probabilistic background, $t=\mu' x$, where x has the distribution (1.1) so that

$$A_q(\kappa) = Et \ , \tag{2.16}$$

$$A_q'(\kappa) = Et^2 - (Et)^2 = \text{var } t \ , \tag{2.17}$$

so that (2.6) and (2.15) are statistically obvious. For $A_q'(\kappa) = \text{var } t$ will be a maximum when $\kappa=0$ because of (2.18) below. It is then equal to $1/q$ by a direct easy calculation. Further, since

$$A_q''(\kappa) = \frac{a_q'''}{a_q} - 3\,\frac{a_q'' a_q'}{a_q^2} + 2\left(\frac{a_q'}{a_q}\right)^3 \ ,$$

$$= Et^3 - 3\,Et^2 Et + 2(Et)^3 \ ,$$

$$= E(t-Et)^3 \ ,$$

so from the evident skewness of the distribution of t for $\kappa > 0$, it follows that

$$A_q''(\kappa) \leqslant 0 \ . \tag{2.18}$$

Thus the function $A_q(\kappa)$ is non-decreasing and convex on $(0,\infty)$, taking its minimum at $\kappa=0$ and maximum as $\kappa\to\infty$ in the range $[0,1]$. While we will not want $A_q(\kappa)$ for negative κ, it is helpful to observe that $A_q(\kappa)$ is an odd function of κ. It is easy to show that A_q, A_q', A_q'' ,... are the successive cumulants of $t=\mu' x$, a fact which can be used to derive the Edgeworth expansion for $\mu' X$.

Integrating a_q'' by parts and using (2.3), we find that $A_q(\kappa)$ satisfies the Riccati equation

$$A_q'(\kappa) = 1 - A_q^2(\kappa) - \frac{q-1}{\kappa} A_q(\kappa) \ , \tag{2.19}$$

as found via Bessel functions by Schou (1978). Similar manipulations show that

$$A_q(\kappa) = \frac{A}{A_{q-2}(\kappa)} - \frac{q-2}{\kappa} \ , \ (q \geq 3) \ , \tag{2.20}$$

as in Schou (1978, A1). By (2.6), (2.15), and (2.8), it follows that

$$A_q'(\infty) = 0 \ , \ A_q(\infty) = 1 \ . \tag{2.21}$$

Inspection of $A_q(\kappa)$ as $\kappa \to 0$ shows $A_q(0) = 0$, so

$$A_q'(\infty) = \frac{1}{q} \ , \ A_q(0) = 0 \ . \tag{2.22}$$

We now give more detailed expansions for $A_q(\kappa)$ as $\kappa \to 0$ and $\kappa \to \infty$ by solving (2.19). The difference equation will be used later to suggest how to tabulate $A_q(\kappa)$ for various values of q.

For $q=3$, there is a simple explicit formula :

$$A_3(\kappa) = \coth \kappa - \kappa^{-1} \ . \tag{2.23}$$

For $q=2$, we may write

$$A_2(\kappa) = I_1(\kappa)/I_0(\kappa) \ , \tag{2.24}$$

since generally

$$A_q(\kappa) = I_{q/2+1}(\kappa)/I_{q/2-1}(\kappa) \ , \tag{2.24'}$$

For $q=2$, the zeros of I_0 nearest to the origin are at $\pm 2.41\,i$ (approx.). For $q=3$, the zeros of $I_{1/2}$ nearest to the origin are at $\pm 3.81\,i$ (approx.). These comments help to define the radius of convergence of the series expansion (2.25) below.

For $\kappa \to 0$, we may substitute $a_1\kappa + a_3\kappa^3 + a_5\kappa^5 + \ldots$ for the (odd) function $A_q(\kappa)$ in (2.19). We then find that

$$A_q(\kappa) = \frac{1}{q}\kappa - \frac{1}{q^2(q+2)}\kappa^3 + \frac{2}{q^3(q+2)(q+4)}\kappa^5 + O(\kappa^7), \tag{2.25}$$

where the first two terms agree with Schou (1978).

For $\kappa \to \infty$, we may substitute a series in powers of κ^{-1} in (2.19) and find that

$$A_q(\kappa) = 1 - \frac{q-1}{2}\frac{1}{\kappa} + \frac{(q-1)(q-3)}{8}\frac{1}{\kappa^2} + \frac{(q-1)(q-3)}{8}\frac{1}{\kappa^3} + O\left(\frac{1}{\kappa^4}\right), \tag{2.26}$$

where the first three terms agree with Schou (1978). For $q=3$, the terms in κ^{-2} and κ^{-3} are zero. But from (2.23), we may write

$$A_3(\kappa) = \left(1 - \frac{1}{\kappa} + \frac{\kappa+1}{\kappa}e^{-2\kappa}\right)(1 - e^{-2\kappa})^{-1}, \tag{2.27}$$

so we see that $A_3(\kappa) - 1 + \frac{1}{\kappa}$ is exponentially small.

From the form of (1.1), it is obvious that maximum likelihood leads to equations for κ of the form $y = A(\kappa)$, so that we need $\kappa = A_q^{-1}(y)$ for all y in $[0,1]$. The equation (2.19) can be rewritten

$$\{(1-y^2)\kappa - (q-1)y\}\frac{d\kappa}{dy} = \kappa . \tag{2.28}$$

Since the inverse function will also be odd, we may put $\kappa = a_1y + a_3y^3 + \ldots$ in (2.28) for y small. We find then, as $y \to 0$,

$$\kappa = q\,y + \frac{q^2}{(q+2)}\,y^3 + \frac{q^3(q+8)}{(q+2)(q+4)}\,y^5 + O(y^7)\;. \tag{2.29}$$

Alternatively, we could revert the series (2.25) to find (2.29). To find κ when y is near unity, we may revert (2.26) using $z = 1-y = 1-A_q(\kappa)$, or set $z = 1-y$ in (2.19) and solve for a series in inverse powers of z. Either way, we find that the solution of $y = A_q(\kappa)$ for y near unity is defined by

$$\frac{1}{\kappa} = \frac{2}{q-1}\,(1-y) + \frac{q-3}{(q-1)}\,(1-y)^2 + \frac{q-3}{(q-1)^2}\,(1-y)^3 + O((1-y)^4)\;. \tag{2.30}$$

The case $q=3$ is again special, but (2.27) yields

$$\frac{1}{\kappa} \sim (1-y) + 2e^{-\kappa}\;,$$

so

$$\frac{1}{\kappa} \sim (1-y) + 2\,\exp - (1-y)^{-1}\;. \tag{2.31}$$

With $v = q/_2-1$, Saw (1981) pointed out that (2.24') can be written as a continued fraction

$$A_q(\kappa) = \frac{\kappa}{2(v+1)+}\;\frac{\kappa^2}{2(v+2)+}\;\frac{\kappa^2}{2(v+3)+}\;\cdots \tag{2.32}$$

and that (2.32) is computationally very convenient. (2.32) is an immediate consequence of the recurrence relation

$$I_{\nu-1}(z) - I_{\nu+1})z) = \frac{2\nu}{z}\,I_\nu(z)\;. \tag{2.33}$$

3. <u>The functions $b_q(\kappa)$, $B_q(\kappa)$ and its inverse.</u>

Here we consider the functions

$$b_q(\kappa) = z_\nu(\kappa) = \int_{-1}^{1} e^{\kappa t^2}(1-t^2)^\nu dt \ , \tag{3.1}$$

$$B_q(\kappa) = N_\nu(\kappa) = b_q'/b_q = z_\nu'/z_\nu \ , \tag{3.2}$$

where $\nu=(q-3)/2 \geq -\frac{1}{2}$ and $-\infty < \kappa < \infty$.

The functions $z_\nu(\kappa)$, $b_q(\kappa)$, and their successive derivatives form descending sequences because $|t| \leq 1$. Further,

$$0 \leq b_{q_1} < b_{q_2} \quad \text{if} \quad q_1 > q_2 \ ,$$

$$0 \leq z_{\nu_1} < z_{\nu_2} \quad \text{if} \quad \nu_1 > q_2 \ .$$

Thus

$$0 \leq N_\nu(\kappa) = B_q(\kappa) \leq 1 \ . \tag{3.3}$$

Elementary calculations show that

$$z_\nu' = z_\nu - z_{\nu+1} \ , \quad (\nu \geq -\tfrac{1}{2}) \ , \tag{3.4}$$

$$2\kappa\, z_\nu' = 2\nu\, z_{\nu-1} - (2\nu+1)z_\nu \ , \ (\nu \geq 0) \ , \tag{3.5}$$

$$2\kappa\, z_\nu'' + (2\nu+3 - 2\kappa)z_\nu' - z_\nu = 0 \ , \ (\nu \geq -\tfrac{1}{2}) \ . \tag{3.6}$$

In this case, we will proceed without identifying and further describing the function $z_\nu(\kappa) = b_q(\kappa)$. Details are given in, e.g., Abramowitz and Stegun (1964).

From Cauchy's inequality and

$$B_q'(\kappa) = b_q''/b_q - (b_q'/b_q)^2 \quad , \tag{3.7}$$

$$b_q''(\kappa) < b_q(\kappa) \quad ,$$

it follows that

$$0 \leqslant B_q'(\kappa) \leqslant 1 - B_q^2(\kappa) \ . \tag{3.8}$$

If the random vector x has the distribution (1.2) and $u = (\mu' x)^2$, then $1 \geqslant Eu \geqslant Eu^2 \geqslant \ldots \geqslant 0$ and

$$B_q(\kappa) = Eu \quad , \tag{3.9}$$

$$B_q'(\kappa) = Eu^2 - (Eu)^2 = \text{var } u \ , \tag{3.10}$$

$$B_q''(\kappa) = Eu^3 - Eu^2Eu - 2Eu(Eu^2-(Eu)^2) = E(u-Eu)^3 \quad . \tag{3.11}$$

From the skewness of the distribution of u, it follows that

$$B_q''(\kappa) \geqslant 0 \quad (\kappa \leqslant 0) \ , \qquad B_q''(\kappa) \leqslant 0 \quad (\kappa \geqslant 0) \ . \tag{3.12}$$

Thus, $B_q(\kappa)$ is nondecreasing and concave on $(-\infty,0)$, convex on $(0,\infty)$, taking its minimum at $-\infty$, and its maximum at $+\infty$ in the range $[0,1]$.

If we put $\kappa=0$ in (3.4), and in (3.5) with $\nu+1$ instead of ν , we find that

$$B_q(0) = 1/q \tag{3.13}$$

Then (3.4) shows that

$$B_q'(0) = 2/q(q+2) \ . \tag{3.14}$$

From (3.4) and (3.5), we find the recurrence relation

$$B_q(\kappa)(1-B_{q-2}(\kappa)) = -\frac{1}{2\kappa} + \frac{q-2}{2\kappa} B_{q-2}(\kappa) . \tag{3.15}$$

Now $B_q(\kappa)$ and $B_{q-2}(\kappa)$ tend to limits (the same from (3.4)) in $[\frac{1}{q}, 1]$ as $\kappa \to \infty$, which (3.15) shows to be

$$B_q(\infty) = 1 . \tag{3.16}$$

Similarly, $B_q(\kappa)$ lies in $[0, \frac{1}{q}]$ for $\kappa \leqslant 0$, and

$$B_q(-\infty) = 0 . \tag{3.17}$$

Finally, we note that $B_q(\kappa)$ satisfies the Riccati equation

$$1 - 2\kappa B_q'(\kappa) - 2\kappa B_q^2(\kappa) = (q-2\kappa)B_q(\kappa) , \tag{3.18}$$

which holds for $q \geqslant 2$.

To find expansions for $B_q(\kappa)$ as $\kappa \to 0$, $\kappa \to +\infty$, and $\kappa \to -\infty$, we may here use either the difference equation (3.15) or the differential equation (3.18). We know of no explicit forms for special values of q, but we will discuss the cases $q=2$ and $q=3$ separately. After harder calculations than the last section, we find:

for $\kappa \to 0$,

$$B_q(\kappa) = \frac{1}{q} + \frac{2(q-1)}{q^2(q+2)} \kappa + \frac{4(q-1)(q-2)}{q^3(q+2)(q+4)} \kappa^2 + \{ \frac{-8(q-1)^2}{q^4(q+2)(q+6)} + \frac{8(q-1)(q-2)^2}{q^4(q+2)(q+4)(q+6)} \} \kappa^3 + O(\kappa^4) ; \tag{3.19}$$

for $\kappa \to +\infty$,

$$B_q(\kappa) = 1 - \frac{q-1}{2}\frac{1}{\kappa} - \frac{q-1}{4}\frac{1}{\kappa^2} - \frac{(q-1)(q+2)}{8}\frac{1}{\kappa^3} + O(\frac{1}{\kappa^4}) \; ; \tag{3.20}$$

for $\kappa \to -\infty$,

$$B_q(\kappa) = -\frac{1}{2}\frac{1}{\kappa} - \frac{(q-3)}{4}\frac{1}{\kappa^2} - \frac{(q-3)(q-4)}{8}\frac{1}{\kappa^3} + O(\frac{1}{\kappa^4}) \; . \tag{3.21}$$

Finally, we need expansions for the solution of $\kappa = B_q(y)$ for y near zero, near $1/q$, and near unity. We may revert the appropriate expansions above, or rewrite (3.18) as

$$\{1 - 2\kappa y^2 + (2\kappa - q)y\} \frac{d\kappa}{dy} = 2\kappa \; . \tag{3.22}$$

For a solution when y is near q^{-1}, the variable needs to be changed to $z = y - q^{-1}$. We find

$$\kappa = \frac{q^2(q+2)}{2}(y - \frac{1}{q}) - \frac{(q-2)(q+2)^2}{2q(q+4)}(y - \frac{1}{q})^2 + O(y - \frac{1}{q})^3) \; . \tag{3.23}$$

As $y \to 1$, $\kappa \to \infty$ and we have

$$\frac{1}{\kappa} = \frac{2}{(q-1)}(1-y) - \frac{2}{(q-1)^2}(1-y)^2 - \frac{4(q+1)}{(q-1)^3}(1-y)^3 + O((1-y)^4) \; . \tag{3.24}$$

As $y \to 0$, $\kappa \to -\infty$ and we find

$$-\frac{1}{\kappa} = 2y + 2(q-3)y^2 - \frac{(q-3)(5q-19)}{4}y^3 + O(y^4) \; . \tag{3.25}$$

The formulas (3.21), (3.26) are unhelpful when $q=3$.

For $q=2$, the density (1.2) is proportional to $\exp \kappa \cos^2\theta$, and has normalizing constant $\int_0^{2} \exp \kappa \cos^2\theta \, d\theta$; when $\kappa > 0$, the modes are at $\theta=0,\pi$, while if $\kappa < 0$, they are at $\pi/2$, $3\pi/2$. Thus there is no need to use $\kappa < 0$ -- a rotation of $\pi/2$ will serve. Further, since $\cos^2\theta = \frac{1}{2}(\cos 2\theta + 1)$, $\phi = 2\theta$ has the density (1.1) with a concentration of $\kappa/2$. Thus,

$$B_2(\kappa) = \tfrac{1}{2}(1 + A_2(\tfrac{\kappa}{2})) . \tag{3.26}$$

It is easily verified that this relation is consistent with A_2 satisfying (2.19) and B_2 satisfying (3.18), etc.. Thus we may check (3.19) against (2.25) and (3.20) against (2.26). In so doing, we find that

$$B_2(\kappa) = \frac{1}{2} + \frac{\kappa}{8} - \frac{1}{16^2}\kappa^3 + O(\kappa^5) . \tag{3.27}$$

Our κ^3 term in (3.19) was so complicated that we hesitated to give it, but this device provides a check of it.

For $q=3$,

$$B_3(\kappa) = \int_{-1}^{1} t^2 e^{\kappa t^2} dt \Big/ \int_{-1}^{1} e^{\kappa t^2} dt .$$

If $\kappa \to -\infty$, set $\kappa = -\lambda$, $\lambda \to \infty$. Then setting $u = (2\lambda)^{\frac{1}{2}} t$,

$$2\lambda \, B(-\lambda) \sim 1 - \frac{2(2\lambda)^{\frac{1}{2}} e^{-\lambda}}{\sqrt{2\pi}} \left\{ 1 - \frac{2}{(2\lambda)^{\frac{1}{2}}} \frac{1}{\sqrt{2\pi}} e^{-\lambda} \right\}^{-1} \tag{3.28}$$

by noting the relation to the Gaussian and the fact that

$$1 - \int_{-\infty}^{x} e^{-z^2/2} \frac{dz}{\sqrt{2\pi}} \sim \frac{x}{\sqrt{2\pi}} e^{-x^2/2} , \quad x \to \infty .$$

Thus (3.28) explains (3.21), i.e., $B_3(\kappa) + (2\kappa)^{-1}$ is exponentially small as $\kappa \to -\infty$. Similarly, as $\kappa \to -\infty$, $\kappa^{-1} + 2y$ (see (3.25)) is exponentially small.

4. The computation of $A_q(\kappa)$, $B_q(\kappa)$ and associated functions.

From (2.10) it follows, that

$$A_q(\kappa) = I_{q/2}(\kappa) \,/\, I_{q/2-1}(\kappa) \quad . \tag{4.1}$$

Since these modified Bessel functions are tabulated (see, e.g., Abramowitz and Stegun (1964)), $A_q(\kappa)$ may be found for any q and κ. To solve $y=A_q(\kappa)$, we may use Newton's method since (2.19),

$$A_q'(\kappa) = 1 - A_q^2(\kappa) - \frac{q-1}{\kappa} A_q(\kappa) \quad , \tag{4.2}$$

and (4.1) enable us to compute the derivative. Thus hand calculations are not a problem for $A_q(\kappa)$.

To produce tables of the $A_q(\kappa)$, machine calculations are needed. If q is odd, we may first compute

$$A_3(\kappa) = \coth \kappa - \kappa^{-1} \; , \tag{4.3}$$

and then use the recurrence relation (2.20) so that

$$A_5(\kappa) = (A_3(\kappa))^{-1} - 3\kappa^{-1} \text{ , etc.}$$

If q is odd, we need separate algorithms for $A_2(\kappa)$ and $A_4(\kappa)$ before (2.20) may be used.

To compute $A_2(\kappa)$ and $A_4(\kappa)$, one may return to the basic definition and use numerical integration to evaluate $a_q(\kappa)$, $a_q'(\kappa)$ for smaller values of κ until the asymptotic expansion (2.26) comes into force and the ratio becomes awkward to handle. The latter will not happen if $a_q(\kappa)$ and $a_q'(\kappa)$ are divided by the first terms of their asymptotic expansions. From (2.10) and the expansion (Watson (1952)) of $I_\nu(\kappa)$, $\kappa \to \infty$, we have

$$a_q(\kappa) \sim (2/\kappa)^{q/2-1}\, e^{\kappa} . \tag{4.4}$$

The derivative of (4.4) will be used for $a_q'(\kappa)$.

If $A_q(\kappa)$ is to be tabulated for a specific q, it may be best to integrate the differential equation (4.2) numerically. To start the solution at $\kappa=0$, set

$$A_q(\kappa) = \kappa/q + \kappa a(\kappa) , \tag{4.5}$$

so that $a(\kappa)$ satisfies

$$a' = -q^{a} - \frac{\kappa}{q^2} - \frac{2\kappa a}{q} - \kappa a^2 , \tag{4.6}$$

which is well behaved at the origin. Of course, $a(0)=0$. When κ becomes large, (4.2) should be used directly.

In Chapter 4 it was shown that the variance stabilizing transformation for κ-estimates is

$$g(\kappa) = \int^{\kappa} (A'(k))^{\frac{1}{2}}\, dk , \tag{4.7}$$

in which the lower terminal is arbitrary. For q=3, we have found it convenient to start at unity. The table of $g(\kappa)$ is then found by numerical integration using (4.2) and (4.3); it is more accurate for values of $\kappa > 1$ than for $\kappa < 1$. Since κ's < 1 are rare for q=3 in the applications we have met, this is very satisfactory. If one regularly dealt with quite large κ's, as in palaeomagnetism, it would be satisfactory to tabulate (4.7) by numerical integration with $A_q'(\kappa)$ replaced by the derivative of the asymptotic expansion (2.25) and starting at κ = 10 or 20 .

Turning now to the computation of $B_q(\kappa)$, it is shown in the Appendix that

$$B_q(\kappa) = \frac{M(3/2,(q+2)/2,\kappa)}{M(1/2,q/2,\kappa)} , \tag{4.8}$$

where $M(a,b,z)$ is one of the Kummer functions. The tabulation most useful for our purposes is that in Rushton and Lang (1954). It gives both the numerator and denominator of (4.8) for $q=1\ (1)\ 5$, $\kappa = 0.02\ (\cdot 02)\cdot 1(\cdot 1)1(1)10(10)50,100,200$. These tables may be extended to negative κ by using Kummer's transformation (5.7).

The only known case where (4.8) reduces to more familiar functions is $q=3$, $\kappa<0$. By specializing (4.8) or a direct calculation,

$$b_3(\kappa) = \kappa^{-\frac{1}{2}} \int_0^{\kappa} e^{u}\, u^{-\frac{1}{2}}\, du \qquad (\kappa>0) , \tag{4.9}$$

and, setting for $\kappa<0$, $-\kappa=\lambda$,

$$b_3(\kappa) = \lambda^{-\frac{1}{2}} \int_0^{\lambda} e^{-u}\, u^{-\frac{1}{2}}\, du \qquad (\kappa<0) . \tag{4.10}$$

Thus, when $\kappa<0$, $b_3(\kappa)$ and $b_3'(\kappa)$ can be found using tables of the incomplete gamma-function.

To compute $B_q(\kappa)$ for specific $q\neq 3$ and κ , or to make tables, it seems that the best method is always to integrate the differential equation (3.18). It may be rewritten as

$$B_q'(\kappa) = -\frac{1}{2\kappa} - B_q^2(\kappa) - \left(\frac{q}{2\kappa} - 1\right) B_q(\kappa) . \tag{4.11}$$

To start the solution, it is essential to use (3.19) since (4.11) is badly behaved for small κ .

To solve $y = B_q(\kappa)$, when none of the expansions (3.19), (3.20), and (3.21) may be used, one must compute $B_q(\kappa)$ and $B_q'(\kappa)$ from (4.11).

To find the variance stabilizing transformation

$$h(\kappa) = \int^{\kappa} (B_q'(k))^{\frac{1}{2}} \, dk \,, \tag{4.12}$$

the range of integration should avoid k=0, if possible.

5. Use of the Kummer Function

It was shown in Section 3 that $b_q(\kappa)$ satisfied the differential equation (3.6), or

$$\kappa\, b_q''(\kappa) + (\tfrac{q}{2} - \kappa)\, b_q'(\kappa) - \tfrac{1}{2}\, b_q(\kappa) = 0 \, . \tag{5.1}$$

This is Kummer's differential equation -- see, e.g., Chapter 13 by L.J. Slater in Abramowitz and Stegun (1964). It is self-adjoint and a confluent form of the hypergeometric equation with a regular singularity at $\kappa=0$ and an irregular singularity at $\kappa=\infty$. All regular solutions at the origin of

$$z\omega'' + (b-z)\omega' - a\omega = 0 \tag{5.2}$$

are proportional to

$$M(a,b,z) = \sum_{r=0}^{\infty} \frac{(a)_r}{(b)_r} \frac{z^r}{r!} \, , \tag{5.3}$$

which arose in the form $M(\frac{q}{2}, \frac{s}{2}, \kappa)$ in the generalized Scheiddegger-Watson distribution discussed in Chapter 5.

Since

$$b_q(\kappa) = \int_{-1}^{1} e^{\kappa t^2} (1-t^2)^{(q-3)/2}\, dt \, ,$$

the results of Section 1 show that

$$\begin{aligned} b_q(0) &= \omega_q/\omega_{q-1} \, , \\ &= \pi^{\frac{1}{2}} \frac{\Gamma((q-1)2)}{\Gamma(q/2)} \, . \end{aligned} \tag{5.4}$$

Thus

$$b_q(\kappa) = \frac{\pi^{\frac{1}{2}}\Gamma((q-1)/2)}{\Gamma(q/2)} M(\tfrac{1}{2}, \tfrac{q}{2}, \kappa) . \tag{5.5}$$

We have already seen a reflection of some of the many recurrence relations satisfied by $M(a,b,z)$, but not the differential property,

$$\frac{d^n}{dz^n} M(a,b,z) = M(a+n , b+n , z) , \tag{5.6}$$

$$M(a,b,z) = e^z M(b-a , b , -z) , \tag{5.7}$$

$$M(1+a-b , 2-b , z) = e^z M(1-a , 2-b , -z) . \tag{5.8}$$

The Kummer function M is related to the Bessel function I_ν, which arose in Section 2 by the formula

$$M(a,b,z) = e^{z/2} \Gamma(b-a-\tfrac{1}{2})(z/4)^{a-b+\frac{1}{2}} \sum_{r=0}^{\infty} \frac{(2b-2a-1)_r(b-2a)_r(-1)^r}{r!\,(b)_r} I_{b-a+\frac{1}{2}+r}(z/2) . \tag{5.9}$$

For $|z|$ large ,

$$\frac{M(a,b,z)}{\Gamma(b)} = \frac{e^{\pm i\pi a}z^{-a}}{\Gamma(b-a)} \{ \sum_{r=0}^{R-1} \frac{(a)_r(1+a-b)_r}{r!} (-z)^{-r} + O(|z|^{-R})\} ,$$
$$+ \frac{e^z z^{a-b}}{\Gamma(a)} \{ \sum_{s=0}^{S-1} \frac{(b-a)_s(1-a)_s}{s!} z^{-s} + O(|z|^{-S})\} \tag{5.10}$$

where the upper sign is taken if $-\pi/2 < \arg z < 3\pi/2$, and the lower otherwise.

From (5.5) and (5.6), it follows that

$$B_q(\kappa) = b_q'(\kappa) \,/\, b_q(\kappa)$$

$$= \frac{d}{d\kappa} \log M(\tfrac{1}{2}\,,\, q/2\,,\, \kappa) \tag{5.11}$$

$$= \frac{M(3/2,(q+2)/2,\kappa)}{M(1/2\,,\, q/2\,,\, \kappa)} \quad .$$

From (5.11) and (5.7), we see that

$$B_q(-\kappa) = \frac{M((q-1)/2,(q+2)/2,\kappa)}{M((q-1)/2,q/2,\kappa)} \quad , \tag{5.12}$$

and (5.11) and (5.8) give yet another form for $B_q(-\kappa)$. (5.5) and (5.9) show that

$$b_q(\kappa) = \pi^{\frac{1}{2}} \frac{\Gamma((q-1)/2)\Gamma((q-2)/2)}{\Gamma(q/2)} \; e^{\kappa/2} \left(\frac{\kappa}{4}\right)^{-(\frac{q-2}{2})}$$

$$\times \sum_{r=0}^{\infty} \frac{(q-2)_r((q-2)/2)_r(-1)^r}{r!\;(q/2)_r} \; I_{q/2+r}\,(\kappa/2)\;; \tag{5.13}$$

(5.5) and (5.10) show that, as $\kappa \to +\infty$,

$$b_q(\kappa) = \pi^{\frac{1}{2}}\,\Gamma((q-1)/2)\; e^{\kappa}\; \kappa^{-(q-1)/2} \tag{5.14}$$

$$\left\{ \sum_{s=0}^{S-1} \frac{((q-1)/2)_s(1/2)_s}{s!} \;\kappa^{-s} + O(\kappa^{-S})\right\} ,$$

while if $\kappa \to -\infty$,

$$b_q(\kappa) = \pi^{\frac{1}{2}} \Gamma((q-1)/2)(-\kappa)^{-\frac{1}{2}}$$

$$\{ \sum_{r=0}^{R-1} \frac{(\frac{1}{2})_r(-(q-1)/2)_r}{r!} (-\kappa)^{-r} + O(\kappa^{-R}) \} \tag{5.15}$$

The expansions (5.13), (5.14), and (5.15) could be used directly to obtain the related expansions for $B_q(\kappa)$. However the procedure used in Section 3 is now seen to be much simpler e.g. it does not require this section.

The maximum likelihood equation in general (see (5.5.3)) requires the inversion of the function

$$\frac{d}{d\kappa} \log M(a,b,\kappa) = \frac{M(a+1,B+1,\kappa)}{M(a,b,\kappa)} \tag{5.16}$$

The recurrence relation (see Abramowitz & Stegun 13.4.7, (1964))

$$b(1-b+\kappa)M(a,b,\kappa) + b(b-1)M(a-1,b-1,\kappa) = a\kappa\, M(a+1,b+1,\kappa) \tag{5.17}$$

immediately yields a continued fraction expansion for (5.16) by analogy with (2.32). It may be expected that this representation will be computationally useful.

Appendix B Asymptotic Spectral Analysis of Cross-Product Matrices

1. Introduction

T. W. Anderson (1963) derived the asymptotic distribution of the eigenvalues and vectors of the covariance matrix of a sample from a Gaussian distribution. Davis (1977) took his basic method and used it to get some results for the non-Gaussian case. The non-Gaussian case is of interest either because one wants to study the sensitivity of methods to deviations from Gaussianity - see, e.g., Muirhead (1982) - or because one has to deal with other distributions. For example, the distribution of the random vector might be entirely restricted to some manifold embedded in $\mathbb{R}^q$ like the surface of the unit sphere or an hyperboloid of rotation. The case of interest to us is the former.

Kim (1978), at the suggestion of R. J. W. Beran, used results from the book by Kato (1976, 1980) on the perturbation theory of linear operators to find the asymptotic distribution of the eigenvalues of the matrix $M_n = n^{-1} \sum_1^n x_i x_i'$, where the x_i's are independently drawn from a certain distribution on the surface Ω_q of the unit sphere in $\mathbb{R}^q$. Tyler (1979, 1981) also used Kato's method to get results in classical multivariate analysis. But the technique is not well-known, nor immediately evident from Kato's book.

Kato's method calls upon Cauchy's Theorem in Complex Variable Theory. Specifically, consider the integral of $(z - z_0)^p$ anti-clockwise around a simple closed curve C in the complex plane which does not go through z_0,

$$\int_C (z - z_0)^p \, dz ,$$

where p is an integer. Unless $p = -1$, it is always zero. When $p = -1$, it is zero if z_0 is outside C and $2\pi i$ where z_0 is inside C.

The techniques and formulae below have many possible applications. Some are given in Watson (1982a), but most remain to be exploited. In the next section, Kato's method is explained for symmetric non-random matrices, and then applied in Section 3 to covariance matrices.

The key formulae in Section 2 and the results of Section 3 have, of course, been obtained before by direct matrix methods, though these are hard to justify. The Kato method not only gives a better insight, but is easier to do and to extend, e.g., to provide asymptotic expansions.

2. The Key to Kato

If T_0 and T_1 are real symmetric $q \times q$ matrices and x is a small real number,

$$T(x) = T_0 + xT_1 \tag{2.1}$$

can be thought of as a linear perturbation of T_0. Let the spectral representation of the matrix T_0 be

$$T_0 = \sum_1^r \lambda_j P_j \; , \; r \leq q \tag{2.2}$$

where

$$\left.\begin{array}{l}\lambda_1, \ldots, \lambda_r \text{ are distinct real numbers,} \\ P_j' = P_j, \quad P_j P_k = \delta_{jk} P_j, \\ \text{rank } P_j = \text{trace } P_j = q_j, \quad \sum_1^r q_j = q .\end{array}\right\} \qquad (2.3)$$

Thus λ_j is an eigenvalue of T_0 that is repeated q_j times. The invariant subspace V_j associated with λ_j has dimension q_j, and P_j projects orthogonally onto V_j whose direct sum is $\mathbb{R}^q$.

The matrix T(x) may have q distinct eigenvalues, but we would expect these to fall into r clusters about $\lambda_1, \ldots, \lambda_r$ and to condense on $\lambda_1, \ldots, \lambda_r$ as $x \to 0$. Equally, the eigenvectors of T(x) should lead us to the eigen subspaces V_j as $x \to 0$. To show how this happens, define the resolvent of T_0, $R_0(\zeta)$ as

$$R_0(\zeta) = (T_0 - \zeta I_q)^{-1} , \qquad (2.4)$$

where ζ is a complex number. By (2.2), we may write

$$R_0(\zeta) = \sum_1^r (\lambda_j - \zeta)^{-1} P_j . \qquad (2.5)$$

Observe that T_0 and $R_0(\zeta)$ commute.

If C is any contour in the complex plane which does not go through any λ_j, which are points on the real axis, Cauchy's Theorem and (2.5) imply that

$$\frac{1}{2\pi i} \int_C R_0(\zeta) d\zeta = \sum_{j \epsilon C} P_j , \qquad (2.6)$$

where the sum is over the projectors P_j associated with eigenvalues λ_j inside C. The integral of a matrix is the matrix of integrals. Similarly,

$$\frac{1}{2\pi i}\int_C T_0R_0(\zeta)d\zeta = \sum_{j\epsilon C}\lambda_j P_j . \qquad (2.7)$$

We observe that the trace of (2.6) gives the sum of the dimensions of the eigen subspaces associated with λ_j within C. Similarly, the trace of (2.7) gives the sum of the eigenvalues (times their multiplicities) within C.

We now consider the resolvent of T(x):

$$R(x,\zeta) = (T(x) - \zeta I_q)^{-1} = R_0(\zeta)(I_q + xT_1R_0(\zeta))^{-1} . \qquad (2.8)$$

If we apply the results of the previous paragraph to $R(x,\zeta)$, we will get information about the eigenvalues $\lambda(T(x))$ and projectors $P(T(x))$ of T(x). As $x \to 0$, we would expect the values of $\lambda(T(x))$ to condense on the eigenvalues λ_j of T_0 .

To obtain the required formulae, we need to expand (2.8) as a power series. For a $q \times q$ matrix A,

$$(I_q + xA)^{-1} = I_q - xA + x^2A^2 - \dots , \qquad (2.9)$$

where the series is absolutely convergent provided $|x| \, \|A\| < 1$, where $\|A\|$ is a norm of A. Thus we can say that for x sufficiently small,

$$(I_q + xA)^{-1} = I_q - xA + 0(x^2) . \qquad (2.10)$$

Applying (2.10) to (2.8), we have, as $|x| \to 0$,

$$R(x,\zeta) = R_0(\zeta) - x R_0(\zeta)T_1R_0(\zeta) + 0(x^2). \qquad (2.11)$$

Consider now the analogue of (2.6) when C_j is a contour which encloses only the eigenvalue λ_j. Then

$$\frac{1}{2\pi i}\int_{C_j} R(x,\zeta)d\zeta = \frac{1}{2\pi i}\int_{C_j} R_0(\zeta)d\zeta + \frac{x}{2\pi i}\int_{C_j} R_0(\zeta)T_1R_0(\zeta)d\zeta + O(x^2) \,. \tag{2.12}$$

The first term on the right hand side (r.h.s.) of (2.12) is P_j. To find the second term, we observe that, if we use (2.5) twice,

$$R_0(\zeta)T_1R_0(\zeta) = \sum_{k=1}^{r}\sum_{\ell=1}^{r} \frac{P_kT_1P_\ell}{(\lambda_k-\zeta)(\lambda_\ell-\zeta)}$$

$$= \sum_{k=1}^{r} \frac{P_kT_1P_k}{(\lambda_k-\zeta)^2} + \sum_{k<\ell} \frac{(P_kT_1P_\ell + P_\ell T_1P_k)}{\lambda_\ell - \lambda_k}\left(\frac{1}{\lambda_k-\zeta} - \frac{1}{\lambda_\ell-\zeta}\right) \tag{2.13}$$

The contour integral of the first term on the r.h.s. of (2.13) is zero. We get contributions from the second term when k or ℓ equal j, and they add to the symmetric matrix

$$\sum_{k\neq j} \frac{P_kT_1P_j + P_jT_1P_k}{\lambda_k - \lambda_j} \,. \tag{2.14}$$

Observe, for later use, that this matrix has a zero trace because P_jP_k is null. Thus

$$\frac{1}{2\pi i}\int_{C_j} R(x,\zeta)d\zeta = P_j + x\sum_{k\neq j}\frac{P_kT_1P_j+P_jT_1P_k}{\lambda_j - \lambda_k} + O(x^2) \tag{2.15}$$

is the analogue of (2.6).

The analogue of (2.7) is obtained by integrating $T(x)R(x,\zeta)$, which may, using (2.1) and (2.11), be written as

$$T(x)R(x,\zeta) = T_0R_0(\zeta) + x(T_1R_0(\zeta) - T_0R_0(\zeta)T_1R_0(\zeta)) + O(x^2). \tag{2.16}$$

The integral of the first term on the r.h.s. of (2.16) is that in (2.7).

To find the second and third terms, we note that

$$T_1R_0(\zeta) = \sum_{k=1}^{r} (\lambda_k-\zeta)^{-1}T_1P_k , \tag{2.17}$$

$$T_0R_0(\zeta)T_1R_0(\zeta) = \sum_{k=1}^{r} \frac{\lambda_kP_kT_1P_k}{(\lambda_k-\zeta)^2}$$

$$+ \sum_{k<\ell} \frac{\lambda_kP_kT_1P_\ell+\lambda_\ell P_\ell T_1P_k}{(\lambda_\ell-\lambda_k)}\left(\frac{1}{\lambda_k-\zeta} - \frac{1}{\lambda_\ell-\zeta}\right), \tag{2.18}$$

where we have used (2.2) and (2.13). Thus we find that

$$\frac{1}{2\pi i}\int_{C_j} T(x)R(x,\zeta)d\zeta = \lambda_jP_j+x(T_1P_j + \sum_{k\neq j} \frac{\lambda_kP_kT_1P_j+\lambda_jP_jT_1P_k}{\lambda_j - \lambda_k})$$

$$+ 0(x^2) . \tag{2.19}$$

Observe that the trace of the second term in the coefficient of x is zero; (2.19) is the analogue of (2.7).

In the applications we have in mind, T(x) will have q distinct eigenvalues $\lambda_1(x), \ldots, \lambda_q(x)$ and (orthonormal) eigenvectors $v_1(x), \ldots, v_q(x)$, so that

$$T(x) = \sum_1^q \lambda_i(x)v_i(x)v_i(x)' \tag{2.20}$$

is the spectral form for T(x). By using the reasoning that led to (2.6) and (2.7), we may then evaluate the l.h.s. of (2.15) and (2.19). Thus,

$$\frac{1}{2\pi i}\int_{C_j} R(x,\zeta)d\zeta = \Sigma\, v_i(x)v_i(x)' \tag{2.21}$$

and

$$\frac{1}{2\pi i}\int_{C_j} T(x)R(x,\zeta)d\zeta = \Sigma\, \lambda_i(x)v_i(x)v_i(x)' , \qquad (2.22)$$

where both sums are over i such that $\lambda_i(x)$ are points inside the contour C_j.

Since trace $v_i(x)v_i(x)' = v_i'(x)v_i(x) = 1$, taking the trace of both sides of (2.15) and using (2.21) yields

$$\#\ \lambda_i(x) \text{ inside } C_j = q_j + O(x^2) \qquad (2.23)$$

for any contour C_j enclosing λ_j. As $x \to 0$, one could use smaller and smaller contours. Hence as $x \to 0$, the eigenvalues of T(x) form <u>clusters</u> of q_j roots about λ_j $(j = 1, \ldots, r)$ which condense upon λ_j. If we do not take the trace of (2.15) and write

$$\hat{P}_j = \sum_{\lambda_i(x)\subset C_j} v_i(x)v_i(x)' , \qquad (2.24)$$

then (2.15) may be written as

$$\hat{P}_j = P_j + x \sum_{k\neq j} \frac{P_k T_1 P_j + P_j T_1 P_k}{\lambda_j - \lambda_k} + O(x^2) . \qquad (2.25)$$

Taking the trace of (2.19) yields

$$\Sigma\, \lambda_i(x) \text{ within } C_j = q_j\lambda_j + x \text{ trace } T_1P_j + O(x^2) , \qquad (2.26)$$

so dividing through by q_j and calling the l.h.s. $\overline{\lambda}_j$, the arithmetic mean of the jth cluster, we have

$$\bar{\lambda}_j = \lambda_j + \frac{x}{q_j} \text{ trace } T_1P_j + O(x^2) \tag{2.27}$$

The formulae (2.25) and (2.27) are ideal for statistical applications, as will be seen in the next section. We close this section by observing that there is no problem except complexity in getting higher order approximations - one merely takes higher order terms in (2.10). For example, the coefficient of x^2 in $R(x_1\ )$ is $R_0T_1R_0T_1R_0$, so using (2.5) and partial fraction expansions, the contour integral may be evaluated to give a lengthy formula. One then finds that (2.23) may be improved to

$$\# \lambda_i(x) \text{ inside } C_j = q_j + O(x^3)\ . \tag{2.28}$$

In Chapter 5, explicit results are given when $r = 2$.

3. Large Sample Theory of Symmetric Cross-Product Matrices

Let x be a random vector in $\mathbb{R}^q$ with components $x^1, x^2, .., x^q$, and suppose that $Ex^ix^jx^kx^\ell$ exists for all $i, j, k, \ell = 1, .., q$. Let x' denote the transpose of x. Call $Exx' = E[x^ix^j] = M$, a symmetric $q \times q$ matrix with spectral form

$$M = \sum_1^r \lambda_jP_j\ . \tag{3.1}$$

If $x_1, .., x_n$ are independent copies of x, define

$$M_n = n^{-1} \sum_1^n x_ix_i'\ . \tag{3.2}$$

Then $M_n \to M$ by the law of large numbers, and by the multivariate central limit theorem

$$n^{1/2}(M_n - M) \xrightarrow{d} G. \tag{3.3}$$

The q(q+1)/2 functionally independent elements of the symmetric matrix G are jointly Gaussian with zero means and a covariance matrix V whose elements are

$$E\, x^i x^j x^k x^\ell - E(x^i x^j) E(m^k x^\ell), \quad i \leq j,\ k \leq \ell. \tag{3.4}$$

To use the results of Section 2, we may write

$$M_n = M + n^{-1/2} \{n^{1/2}(M_n - M)\}, \tag{3.5}$$

instead of

$$T(x) = T_0 + x\, T_1 .$$

From (3.3), T_1 corresponds to G, x to $n^{-1/2}$, and M to T_0. Provided no λ_j in (3.1) is zero, the matrix M_n will, with probability one, have distinct eigenvalues - Okamoto (1973). If, say, $\lambda_1 = 0$, $E(P_1x)(P_1x)'$ is a matrix of zeros so that, taking the trace, $E\,\|P_1x\|^2 = 0$. Thus P_1x is a null vector and M_n will have q_1 zero roots, and the data will determine the eigen subspace V_1 exactly. This case has little interest, so we assume that all the $\lambda_j > 0$.

The matrix M_n will be used to estimate the λ_j and P_j, j=1, ..., r. Combining (3.5) with (2.24), (2.25), and (2.27), we have the key results:

for j = 1, ..., r,

$$n^{1/2}\,(\hat{P}_j - P_j) \xrightarrow{d} \sum_{k \neq j} \frac{P_k G P_j + P_j G P_k}{\lambda_j - \lambda_k} \tag{3.6}$$

$$n^{1/2}\,(\bar{\lambda}_j - \lambda_j) \xrightarrow{d} \frac{1}{q_j} \operatorname{trace} GP_j . \tag{3.7}$$

The r.h.s.'s of (3.6) and (3.7) are linear in the Gaussian matrix G so that the l.h.s.'s have asymptotically Gaussian distributions with zero means and variances and covariances that depend upon the covariance matrix V of G.

(3.7) is univariate and so easy to understand, e.g., it leads to a normal confidence interval for λ_j, although we will see that one will do better with a transformation. (3.6) describes the difference between estimated and true projectors, and needs further simplification. Using the Euclidean matrix norm ($\|A\|^2$ = trace AA'),

$$n\|\hat{P}_j - P_j\|^2 \xrightarrow{d} 2 \sum_{k\neq j} \frac{\text{trace } P_j G P_k G}{(\lambda_j - \lambda_k)^2} . \tag{3.8}$$

Again, one might examine the different effects of $\hat{P}_j$ and P_j on vectors. For example, if $v \in V_j$,

$$n^{1/2}(\hat{P}_j v - P_j v) \xrightarrow{d} \sum_{k\neq j} \frac{P_k G v}{\lambda_j - \lambda_k} , \tag{3.9}$$

$$n\|\hat{P}_j v - P_j v\|^2 \xrightarrow{d} \sum_{k\neq j} \frac{v' G P_k G v}{\lambda_j - \lambda_k} . \tag{3.10}$$

More fundamentally, if $\hat{V}_j$ is the subspace onto which $\hat{P}_j$ projects, $\hat{V}_j$ will be "close" to V_j if $\cos\theta = v'\hat{v}$ is always large when $v \in V_j$ and $\hat{v} \in \hat{V}_j$, $\|v\| = 1$, $\|\hat{v}\| = 1$. Thus we should seek the stationary values of $(P_j u)'(\hat{P}_j w)$, given $\|P_j u\| = \|\hat{P}_j w\| = 1$, i.e., we should consider

$$2u'P_j\hat{P}_j w - \theta u'P_j u - \phi w'\hat{P}_j w ,$$

where θ and ϕ are Lagrangian multipliers. Hence

$$\left.\begin{aligned} P_j\hat{P}_j w - \theta P_j u &= 0 , \\ \hat{P}_j P_j u - \phi \hat{P}_j w &= 0 , \end{aligned}\right\} \tag{3.11}$$

so that

$$\begin{aligned} \theta = \phi &= \text{stationary value of } (P_j u)'(\hat{P}_j w) \\ &= C, \text{ say .} \end{aligned}$$

Hence the equations (3.11) will only have a solution if

$$\begin{vmatrix} -C\,P_j & P_j\,\hat{P}_j \\ \hat{P}_j\,P_j & -C\,\hat{P}_j \end{vmatrix} = 0 \tag{3.12}$$

This equation for C may be reduced to

$$\left| P_j\,\hat{P}_j\,P_j - C^2\,P_j \right| = 0 , \tag{3.13}$$

which has q_j non-zero roots $C_\ell^{\,2}$. If, however, (3.6) is used, one finds eventually that all the $C_\ell^{\,2}$ are unity. Chapter 5 deals with the case where $r = 2$, and shows, by taking the next term in the expensions in Section 2, that $n(1 - C_\ell^{\,2})$ have asymptotic distributions. It is conjectured that for any r, the asymptotic joint distribution of $n(1-C_1)$, ..., $n(1 - C_{q_j})$ is the joint distribution of the non-zero eigenvalues of

$$\sum_{k \neq j} \frac{P_j G P_k G P_j}{(\lambda_j - \lambda_k)^2} \tag{3.14}$$

Some of the above results become easier to understand if we write, since $I_q = P_1 + \ldots + P_r$,

$$y_j = P_j x, \quad x = y_1 + \ldots + y_r, \tag{3.15}$$

One of the reasons results become simpler for the Gaussian is that there $y_1, \ldots, y_r$ are independent. Since $Exx' = M = \lambda_j P_j$,

$$\left.\begin{array}{l} Ey_k y_\ell' = 0,\ \ell \neq k,\ Ey_j y_j' = \lambda_j P_j \\ Ey_\ell' y_k = 0,\ (\ell \neq k),\ Ey_j' y_j = \lambda_j q_j \end{array}\right\} . \tag{3.16}$$

Thus (3.7) may be rewritten as

$$n^{1/2} (\bar{\lambda}_j - \lambda_j) \sim \frac{n^{1/2}}{q_j} \text{trace} \left(\frac{1}{n} \sum_{i=1}^{n} y_{ji} y_{ji}' - \lambda_j P_j \right)$$

$$= \frac{n^{1/2}}{q_j} \left(\frac{1}{n} \sum_{i=1}^{n} y_{ji}' y_{ji} - \lambda_j q_j \right), \tag{3.17}$$

so that by (3.16) and the Central Limit Theorem

$$\mathcal{L}\ n^{1/2} (\bar{\lambda}_j - \lambda_j) \longrightarrow G_1(0, \text{var}(y_j' y_j)\, q_j^{-2}), \tag{3.18}$$

where $G_q(\mu, \Sigma)$ stands for the Gaussian distribution in q dimensions with mean vector μ and covariance matrix Σ. Similarly, (3.6) can be written as

$$n^{1/2} (\hat{P}_j - P_j) \sim n^{-1/2} \sum_{i=1}^{n} \sum_{k \neq j} \frac{y_{ki} y_{ji}' + y_{ji} y_{ki}'}{\lambda_j - \lambda_k} \tag{3.19}$$

If $\mathcal{L}\, x = G_q(0,M)$, $y_1, y_2, \ldots, y_\gamma$ are independent and $\mathcal{L}\, y_j' y_j \lambda_j^{-1} = \chi^2_{q_j}$ so, that $\text{var}(y_j' y_j) = \lambda_j^2 2q_j$. Then (3.18) reads:

$$\mathcal{L}\, n^{1/2}\, (\bar{\lambda}_j - \lambda_j) \longrightarrow G_1(0,\, 2\lambda_j^2/q_j).$$

Hence

$$\mathcal{L}\, n^{1/2}\, (\log\, (\bar{\lambda}_j/\lambda_j) - 1) \longrightarrow G_1(0,\, 2/q_j), \tag{3.20}$$

giving the variance stabilizing transformation. Moreover, in this Gaussian case, the $n^{1/2}\, (\bar{\lambda}_j - \lambda_j)$ or $n^{1/2}\, (\log(\bar{\lambda}_j/\lambda_j) - 1)$ are asymptotically independent, a simplifying result which is not true in general. Under no circumstances could it be expected that the $\hat{P}_j$ would be independent, since $\sum_{j=1}^{r} \hat{P}_j = I_q$.

With this introduction, the compact paper by Tyler (1981) may be read for more details on $\hat{P}_j$. He also gives tests. For the special case of $r = 2$ and distributions restricted to Ω_q, see Watson (1982a). If an additional assumption is made that the distribution of x depends only upon $\|y_1\|, \ldots, \|y_r\|$, more results may be derived - see Watson(1982b).

4. Direct Approach to Large Sample Theory of Cross-Product Matrices

The eigenvalues of M_n are the roots $\hat{\lambda}$ of

$$|M_n - \lambda I| = |M - \lambda I + \frac{1}{\sqrt{n}} T_1| = 0\ , \tag{4.1}$$

where, as in Section 3, $M = \sum_1^r \lambda_j P_j$, $T_1 = \sqrt{n}(M_n - M)$. Suppose orthonormal eigenvectors are selected to span each of the invariant subspaces V_j and arranged as column vectors to form a $q \times q$ orthogonal matrix H. Let the first q_1 columns correspond to V_1, the next q_2 columns to V_2, etc., and write it in partitioned form

$$H = [H_1, \ldots, H_r] . \tag{4.2}$$

Then since $H'H = HH' = I_q$, we have

$$H_a'H_b = 0(a \neq b) , \quad H_a'H_a = I_{q_a} ,$$

$$H_1H_1' + \ldots + H_rH_r' = I_q , \tag{4.3}$$

$$H_aH_a' = P_a , \; a = 1, \ldots, r ,$$

and

$$H'MH = D(\lambda_j I_{q_j}) , \tag{4.4}$$

a matrix partitioned so all r^2 submatrices are zero except for the multiples of identity matrices on the diagonal.

If H is applied to (4.1), it takes the partitioned form.

$$\left| (\lambda_i - \lambda) I_{qi} \delta_{ij} + n^{-1/2} H_i'T_1H_j \right| = 0 . \tag{4.5}$$

Since $n \to \infty$, we seek the $0(1)$ and $0(n^{-1/2})$ terms only in the expansion of (4.5). Applying the formula

$$\begin{vmatrix} A & C \\ B & D \end{vmatrix} = |A| \, |D - BA^{-1}C|$$

when A is the leading submatrix of (4.5), it is seen that $BA^{-1}C$ is $O(n^{-1})$, and so negligible. Hence we may repeat the procedure to find that equation (4.5), and hence (4.1), is, to this order:

$$\prod_{j=1}^{r} |(\lambda_j - \lambda) I_{q_j} + n^{-1/2} H_j' T_1 H_j| = 0 . \tag{4.6}$$

This shows that the eigenvalues of M_n, for large n, form clusters about the r distinct roots λ_j of M. Expanding the jth factor in (4.6) to $O(n^{-1/2})$, we only need the product of the diagonal terms, and find the equation

$$(\lambda_j - \lambda)^{q_j}(1 + n^{-1/2} \frac{\text{trace } H_j' T_1 H_j}{\lambda_j - \lambda}) = 0 . \tag{4.7}$$

Since trace $H_j' T_1 H_j$ = trace $H_j H_j' T_1$ = trace $P_j T_1$ by (4.3), the q_j roots of (4.7) tend to λ_j as $n \to \infty$, and the leading terms of the polynomial (degree q_j) equation for λ are

$$\lambda^q - \{q\lambda_j + n^{-1/2} \text{ trace } P_j T_1\}\lambda^{q-1} + \ldots = 0 , \tag{4.8}$$

so that if the roots of this equation are denoted by $\hat{\lambda}$,

$$\Sigma\hat{\lambda} = q\lambda_j + n^{-1/2} \text{ trace } P_j T_1 , \tag{4.9}$$

as we found in (3.27). But one cannot expect to obtain the roots in the cluster for λ_j from (4.7) (it gives them to be λ_j(q-1 times), $\lambda_j + n^{-1/2}$ trace $P_j T_1$ (once)) because, when λ is within $n^{-1/2}$ of λ_j all the terms in the matrix in (4.6) are of order $n^{-1/2}$. However, (4.8) does give the correct coefficient for λ^{q-1} in (4.6).

Since the eigenvalues of M_n will, in general be distinct, the approximations made above are inadequate to discuss e.g., the joint distribution of the eigenvalues in a cluster.

The following direct derivation of the analogue of (2.25) or (3.6) is harder to justify. Write

$$\hat{P}_j = P_j + n^{-1/2}\Delta, \quad \Delta = n^{1/2}(\hat{P}_j - P_j) \tag{4.10}$$

and, because the roots in the cluster are within n^{-1} of λ_j, set

$$M_n\hat{P}_j = (\lambda_j + n^{-1/2}\delta)\hat{P}_j \ , \tag{4.11}$$

i.e.,

$$(M + n^{-1/2}T_1)(P_j + n^{-1/2}\Delta) = (\lambda_j + n^{-1/2}\delta)(P_j + n^{-1/2}\Delta),$$

so that the terms in $n^{-1/2}$ yield the equation

$$T_1P_j + M\Delta = \lambda_j\Delta + \delta P_j ,$$

or

$$(M - \lambda_j I)\Delta = -T_1P_j + \delta P_j \ . \tag{4.12}$$

But $M - \lambda_j I = \sum_{k\neq j}(\lambda_k - \lambda_j)P_k$, so that we could replace Δ by $\Delta - P_jX$ and still satisfy (4.12). Thus (4.12) is solved by multiplying (4.12) by $\Sigma(\lambda_k - \lambda_j)^{-1}P_k$ and adding P_jX so that

$$\Delta = \sum_{k\neq j}(\lambda_j - \lambda_k)^{-1}P_kT_1P_j + P_jX.$$

However, from (4.10), we see that Δ must be symmetric, and this determines X. Thus

$$\Delta = n^{1/2}(\hat{P}_j - P_j) = \sum_{k \neq j} \frac{P_k T_1 P_j + P_j T_1 P_k}{\lambda_j - \lambda_k} , \qquad (4.13)$$

which is the desired result. However, without the results of Section 3, (4.10) and (4.11) are merely intuitions.

Appendix C A Series Expansion for the Angular Gaussian Distribution

Christopher Bingham

The so called angular Gaussian distribution is the marginal distribution of $\underset{\sim}{\ell}$ when $x = r\underset{\sim}{\ell}$, $r\underset{\sim}{\ell}$ $r = \|\underset{\sim}{x}\|$, is distributed as $N_q(\underset{\sim}{\mu},\sigma^2 I_q)$. Let $\underset{\sim}{\mu} = m\sigma\underset{\sim}{\lambda}$, where $m = \|\mu\| /\sigma$. Then the joint distribution of r and $\underset{\sim}{\ell}$ is

$$(2\pi)^{-\frac{1}{2}q}\sigma^{-q}r^{q-1}\exp\{ -\tfrac{1}{2}\|r\underset{\sim}{\ell} - m\sigma\underset{\sim}{\lambda}\|^2/\sigma^2 \}\, dr\, d\omega_q(\ell)$$

$$= (2\pi)^{-\frac{1}{2}q}\sigma^{-q}r^{q-1} \exp\{ -\tfrac{1}{2}r^2/\sigma^2 + mr\underset{\sim}{\ell}'\underset{\sim}{\lambda}/\sigma \}\, dr\, d_q(\underset{\sim}{\ell})$$

Substituting $s = r/\sigma$, $dr = \sigma ds$, and integrating out s , we obtain the marginal density of $\underset{\sim}{\ell}$ with respect to $d\omega_q(\underset{\sim}{\ell})/\omega_q$ as

$$\omega_q(2\pi)^{-\frac{1}{2}q}e^{-\frac{1}{2}m^2}\int_0^\infty s^{q-1}e^{-\frac{1}{2}s^2}e^{sm\cos\theta}\,ds \ , \quad \cos\theta = \underset{\sim}{\ell}'\underset{\sim}{\lambda} \ ,$$

$$= \frac{2}{2^{\frac{1}{2}q}\Gamma(\frac{1}{2}q)}e^{-\frac{1}{2}m^2}\int_0^\infty s^{q-1}e^{-\frac{1}{2}s^2}e^{sm\cos\theta}\,ds \ . \qquad (1)$$

Now it is a classical result (Erdelyi et al, Vol. 2, p. 98, eq. (1)) that

$$e^{q\cos\theta} = (\tfrac{1}{2}z)^{-\nu}\Gamma(\nu)\sum_{k=0}^{\infty}(\nu+k)\, I_{\nu+k}(z)C_k^{\nu}(\cos\theta) \ , \ \nu > 0 \ , \qquad (2)$$

where $C_k^{\nu}(x)$ is a Gegenbauer polynomial (Erdelyi et al, Vol. 2, p. 235) and $I_\alpha(z)$ is a modified Bessel function of the first kind (*ibid*, p. 5). Since (see, *ibid* p. 235))

$$\lim_{\nu\to 0} \Gamma(\nu)C_k^{\nu}(x) = \lim_{\nu\to 0} \nu^{-1}C_k^{\nu}(x) = 2k^{-1}T_k(x) \ , \ k = 1,2,\ldots, \tag{3}$$

$$= 2k^{-1}\cos(k\cos^{-1}x) \ ,$$

when $\nu = 0$, (2) takes the form

$$e^{z\cos\theta} = I_0(z) + 2\sum_{k=1}^{\infty} I_k(z)\cos k\theta . \tag{4}$$

It is convenient to express I_α in generalized hypergeometric function notation, according to which

$${}_rF_t(a_1,\ldots,a_r;b_1,\ldots,b_t;x) = \sum_{j=0}^{\infty} \frac{(a_1)_j\cdots(a_r)_j}{(b_1)_j\cdots(b_t)_j}\frac{x^j}{j!} ,$$

where $(\alpha)_j$ is the ascending factorial

$$(\alpha)_j = \alpha(\alpha+1) \ \ldots \ (\alpha+j-1) = \Gamma(\alpha+j)/\Gamma(\alpha) \ .$$

The modified Bessel functions occurring in (2) and (4) are expressible as

$$I_\alpha(z) = (\tfrac{1}{2}z)^\alpha \ {}_0F_1(\alpha + 1;\ z^2/4)/\Gamma(\alpha+1) \ . \tag{5}$$

Thus (2) can be expressed as

$$e^{z\cos\theta} = \sum_{k=0}^{\infty} (\tfrac{1}{2}z)^k \frac{{}_0F_1(\nu+k+1;z^2/4)}{(\nu)_k} C_k^{\nu}(\cos\theta),\ \nu > 0 , \tag{2'}$$

With a similar form corresponding to (4):

$$e^{z\cos\theta} = {}_0F_1(1;z^2/4) + 2\sum_{k=1}^{\infty}(\tfrac{1}{2}z)^k(k!)^{-1}{}_0F_1(k+1;z^2/4)\cos k\theta \,. \qquad (4')$$

Substituting (2') with $\nu = \frac{1}{2}q-1$ in (1), we have for the marginal density of $\underset{\sim}{\ell}$ with respect to $d\omega_q/\omega_q$,

$$(2^{\frac{1}{2}q-1}\Gamma(\tfrac{1}{2}q))^{-1} e^{-\frac{1}{2}m^2}\sum_{k=0}^{\infty}(\tfrac{1}{2}m)^k \frac{1}{(\frac{1}{2}q-1)_k}\left[\int_0^{\infty} s^{q+k-1} e^{-\frac{1}{2}s^2} {}_0F_1(\tfrac{1}{2}q+k;(\tfrac{1}{2}ms)^2)ds \; C_k^{\frac{1}{2}q-1}(\cos\theta)\right].$$

Consider the integral in (6). Expanding ${}_0F_1$ and integrating term by term, we obtain (using $\int_0^{\infty} e^{-\frac{1}{2}s^2}s^{\gamma-1}ds = 2^{\frac{1}{2}\gamma-1}\Gamma(\frac{1}{2}\gamma)$)

$$\int_0^{\infty} s^{q+k-1}e^{-\frac{1}{2}s^2}\,{}_0F_1(\tfrac{1}{2}q+k;(\tfrac{1}{2}ms)^2)ds = \sum_{j=0}^{\infty}\frac{(\frac{1}{2}m)^{2j}}{(\frac{1}{2}q+k)_j j!}\int_0^{\infty} s^{q+k+2j-1}\,e^{-\frac{1}{2}s^2}ds$$

$$= \sum_{j=0}^{\infty}\frac{(\frac{1}{2}m)^{2j}}{(\frac{1}{2}q+k)_j j!}\,2^{\frac{1}{2}(q+k)+j-1}\Gamma(\tfrac{1}{2}q+\tfrac{1}{2}k+j)$$

$$= 2^{\frac{1}{2}q+\frac{1}{2}k-1}\Gamma(\tfrac{1}{2}q+\tfrac{1}{2}k)\;{}_1F_1(\tfrac{1}{2}q+\tfrac{1}{2}k;\tfrac{1}{2}q+k;\tfrac{1}{2}m^2)$$

$$= 2^{\frac{1}{2}q+\frac{1}{2}k-1}\Gamma(\tfrac{1}{2}q+\tfrac{1}{2}k)\;e^{\frac{1}{2}m^2}\;{}_1F_1(\tfrac{1}{2}k;\tfrac{1}{2}q+k;-\tfrac{1}{2}m^2)\,, \qquad (7)$$

the last equality following from Kummer's identity (Erdelyi, Vol. I, p. 253). The ${}_1F_1$ function, is a confluent hypergeometric function, sometimes called Kummer's function and is sometimes denoted by

$${}_1F_1(a;c;z) = M(a,\, c,\, z) = \Phi(a,\, c,\, z)\,.$$

Note that ${}_1F_1(0;c;z) \equiv 1$. Thus the marginal density of $\underset{\sim}{\ell}$ with respect to $d\omega_q/\omega_q$ is

$$1 + \sum_{k=1}^{\infty} \frac{2^{\frac{1}{2}k}\Gamma(\tfrac{1}{2}q+\tfrac{1}{2}k)\Gamma(\tfrac{1}{2}q-1)}{\Gamma(\tfrac{1}{2}q)\ (\tfrac{1}{2}q+k-1)} (\tfrac{1}{2}m)^k \ {}_1F_1(\tfrac{1}{2}k;\tfrac{1}{2}q+k;-\tfrac{1}{2}m^2)C_k^{\frac{1}{2}q-1}(\cos\theta),\ q=3,4,\ldots$$

$$= 1 + (\tfrac{1}{2}q-1)^{-1} \sum_{k=1}^{\infty} \frac{\Gamma(\tfrac{1}{2}q+\tfrac{1}{2}k)}{\Gamma(\tfrac{1}{2}q+k-1)} (m^2/2)^{\frac{1}{2}k}\ {}_1F_1(\tfrac{1}{2}k;\tfrac{1}{2}q+k;-\tfrac{1}{2}m^2)\ C_k^{\frac{1}{2}q-1}(\cos\theta)\ . \qquad (8)$$

When $q=2$, the corresponding result is

$$1 + 2 \sum_{k=1}^{\infty} \Gamma(\tfrac{1}{2}k+1)(k!)^{-1}\ (\tfrac{1}{2}m^2)^{\frac{1}{2}k}\ {}_1F_1(\tfrac{1}{2}k;k+1;-\tfrac{1}{2}m^2)\cos k\theta\ . \qquad (9)$$

Equations (8) and (9) should be compared with corresponding expansions for the Langevin density relative to $d\omega_q/\omega_q$ obtainable from (2'):

$$\frac{\exp(\kappa\cos\theta)}{{}_0F_1(\tfrac{1}{2}q;\tfrac{1}{4}\kappa^2)} = 1 + \sum_{k=1}^{\infty} (\tfrac{1}{2}\ )^k \frac{{}_0F_1(\tfrac{1}{2}q+k;\kappa^2/4)}{(\tfrac{1}{2}q-1)_k\,{}_0F_1(\tfrac{1}{2}q;\kappa^2/4)}\ C_k^{\frac{1}{2}q-1}(\cos\theta)\ . \qquad (10)$$

The coefficients of $(\tfrac{1}{2}q-1)^{-1}C_1^{\frac{1}{2}q-1}(\cos\theta)$ in (8) and (10) are, respectively,

$$\Gamma(\tfrac{1}{2}q+\tfrac{1}{2})(\Gamma(\tfrac{1}{2}q))^{-1}(\tfrac{1}{2}m^2)^{\frac{1}{2}}\ {}_1F_1(\tfrac{1}{2};\tfrac{1}{2}q+1;\ -\ \tfrac{1}{2}m^2) = \frac{\Gamma(\tfrac{1}{2}q+\tfrac{1}{2})}{\Gamma(\tfrac{1}{2}q)}\ 2^{-\frac{1}{2}}m(1+O(m^2))\ ,$$

and

$$(\tfrac{1}{2}\kappa)\ {}_0F_1(\tfrac{1}{2}q+1;\kappa^2/4)/{}_0F_1(\tfrac{1}{2}q;\kappa^2/4) = \tfrac{1}{2}\kappa(1+O(\kappa^2))\ .$$

For small values of m (low concentration), one can approximately match these coefficients of lowest order by taking

$$\kappa \cong 2^{\frac{1}{2}} \frac{\Gamma(\frac{1}{2}q+\frac{1}{2})}{\Gamma(\frac{1}{2}q)} m . \tag{11}$$

For $q=2$, this yields $\kappa \cong 2^{\frac{1}{2}}\Gamma(3/2)m = (\frac{1}{2}\pi)^{\frac{1}{2}}m$, and for $q=3$ we get $\kappa \cong (2^{\frac{1}{2}}\Gamma(2)/\Gamma(3/2))m = (8/\pi)^{\frac{1}{2}}m$.

For large m (high concentration), we have (see Erdelyi, Vol. 2, p. 86, eq. (5))

$$\frac{(\frac{1}{2}\kappa)\,{}_0F_1(\frac{1}{2}q+1;\kappa^2/4)}{{}_0F_1(\frac{1}{2}q;\kappa^2/4)} = \frac{\Gamma(\frac{1}{2}q+1)}{\Gamma(\frac{1}{2}q)} \frac{I_{\frac{1}{2}q}(\kappa)}{I_{\frac{1}{2}q-1}(\kappa)} = \tfrac{1}{2}q \frac{1 + (1-q^2)(8\kappa)^{-1} + O(\kappa^{-2})}{1 + (1-(q-2)^2)(8\kappa)^{-1} + O(\kappa^{-2})}$$

$$= \tfrac{1}{2}q(1 - (q-1)/(2\kappa) + O(\kappa^{-2})) ,$$

and (see Erdelyi, Vol. I, p. 278, eq. (2))

$$\frac{\Gamma(\frac{1}{2}q+\frac{1}{2})}{\Gamma(\frac{1}{2}q)} (\tfrac{1}{2}m^2)^{\frac{1}{2}}\, {}_1F_1(\tfrac{1}{2};\tfrac{1}{2}q+1;-\tfrac{1}{2}m^2) = \frac{\Gamma(\frac{1}{2}q+1)}{\Gamma(\frac{1}{2}q)} (2^{-\frac{1}{2}}m)(2^{\frac{1}{2}}m^{-1})(1 - \frac{\frac{1}{2}(\frac{1}{2}q-\frac{1}{2})}{\frac{1}{2}m^2} + O(m^{-4}))$$

$$= \tfrac{1}{2}q(1 - (q-1)/(2m^2) + O(m^{-4})) .$$

Thus, nearly optimal matching of the coefficients of $C_1^{\frac{1}{2}q-1}$ for large concentration is obtained by taking

$$\kappa \cong m^2 . \tag{12}$$

These results are valid for $q=2,3,\ldots$. By further computations, one can show that the choice $\kappa = m^2$ matches all coefficients of $C_k^{\frac{1}{2}q-1}$ through the $O(m^{-2})$ (or $O(\kappa^{-1})$) terms.

References

Abramowitz, M., Stegun, I.A. 1965. Handbook of Mathematical Functions, N.B.S. Applied Mathematics Series .55, U.S. Government Printing Office, Washington, 1046 pp.

Anderson, T.W., 1959. An Introduction to Multivariate Analysis, John Wiley & Sons, New York, 374 pp.

Anderson, T.W.A., Stephens, M.A., 1972. "Tests for randomness of directions against equatorial and bimodal alternatives," Biometrika, 59, 613-621.

Arnold, K.J., 1941. "On Spherical Probability Distributions," Ph.D. Thesis, M.I.T., Boston.

Barndorff-Nielsen, O., 1978. Information and Exponential Families in Statistical Theory, John Wiley & Sons, Chichester & New York, 238 pp.

Batschelet, E., 1971. "Recent statistical methods for orientation," (Animal Orientation Symposium 1970 on Wallops Island), Amer. Inst. Biol. Sciences, Washington, D.C.

Beran, R.J., 1968. "Testing for uniformity on a compact homogeneous space," J. App. Prob. 5, 177-95.

Beran, R.J., 1979. "Exponential models for directional data," Ann. Statist., 7, 1162-1178.

Bernoulli, D., 1734. "Resherches physiques et astronomiques, sur le problème proposé pour la second fois par l'Académie Royale des Sciences des Paris," récuil des pieces qui ont remporté le prix de l'Académie Royale des Sciences, Tome III, 95-134.

Bhattacharya, R.N., Rao, R.R., 1976. Normal Approximation and Asymptotic Expansions, John Wiley & Sons, New York, 274 pp.

Bingham, C., 1964. "Distributions on the sphere and propetive plane," Ph.D. Thesis, Yale University.

Bingham, C., 1974. "An antipodally symmetric distribution on the sphere," Ann. Statist., 2, 1201-1225.

Durand, D., Greenwood, J.A., 1957. "Random unit vectors II: usefulness of Gram-Charlier and related series in approximating distributions," Ann. Math Statist., 28, 978-986.

Dyson, F.J., Lieb, E.H., Simon, B., 1978. "Phase transitions in quantum systems with isotropic and nonisotropic interactions," J. Stat. Phys., 18, 335-383.

Erdelyi, A. et al, 1955. Transcendental Functions, Vols. 1, 2, 3, McGraw-Hill New York.

Feller, W., 1971. An Introduction to Probability Theory and Its Applications, John Wiley & Sons, New York, 669 pp.

Fisher, R.A., 1953. "Dispersion on a sphere," Proc. Roy. Soc. Lond., A217, 295-305.

Fujikoshi, Y., 1980. "Asymptotic expansions for the distributions of sample roots under non-normality," Biometrika, 67 #1, 45-51.

Gordon, L., Hudson, M., 1977. "A characterization of the Von Mises Distribution," Ann. Statist., 5, 813-814.

Gould, J.L., 1981. "Human homings - an elusive phenomenon," Science, 212, 1061-1063.

Greenwood, J.A., Durand, D., 1955. "The distribution of length and components of the sum of n random unit vectors," Ann. Math Statist., 26, 233-246.

Hammersley, J.M., 1950. "On estimating restricted parameters," J.R. Statist. Soc. (B), 12, 192-

Hamming, R.W., 1973. Numerical Methods for Scientists and Engineers, McGraw-Hill, New York, 721 pp.

Hartman, P., Watson, G.S., 1974. "Normal" distribution functions on spheres and the modified Bessel function," Ann. Prob., 2, 593-607.

Hetherington, T.J., 1981. "Analysis of Directional Data by Exponential Models, Ph.D. Thesis, University of California, Berkeley.

Jensen, J.L., 1981. "On the hyperboloid distribution," Scand. J. Statist., 8, 193-206.

Johansen, S., 1979. Introduction to the Theory of the Regular Exponential Families, Lecture Notes No. 3, Institute of Mathematical Statistics, University of Copenhagen, 94 pp.

Kagan, A.M., Linnik Yu K., Rao, C.R., 193 . Characterization Problems in Mathematical Statistics, John Wiley & Sons, New York, 499 pp.

King, M.L., 1980. "Robust tests for spherical symmetry and their application to least squares regression," Ann. Statist., 8, 1265-1271.

Kluyver, J.C., 1906. "A local probability theorem," Ned. Akad. Wet. Proc., A8, 341-350.

Kendall, D.G., 1974. "Pole seeking Brownian motion and bird navigation," Joy. Roy. Stat. Soc. B., 36, 365-417.

Kent, J., 1976. "Distributions, processes and statistics on "spheres," Ph.D. Thesis, Univeristy of Cambridge.

Krumbein, W.C., 1939. "Preferred orientation of pebbles in sedimentary deposits," J. Geol., 47, 673-706.

Langevin, P., 1905. "Magnetisme et theorie des electrons," Ann. de Chim. et de Phys., 5, 70-127.

Lord, R.D., 1948. "A problem with random vectors," Phil. Mag., 39, 66-71.

MacDuffee, C.C., 1946. The Theory of Matrices, Chelsea Pub. Co., New York, 110 pp.

Magnus, J.R., Nendecker, H., 1979. "The commutation matrix: some properties and applications," Ann. Statist., 7, No. 2, 381-394.

Mardia, K.V., 1972. Statistics of Directional Data, Academic Press, New York, 357 pp.

Mardia, K.V., 1975. "Characterization of directional distributions," (in Statistical Distributions) in Scientific Work, G.P. Patil et al (Eds.) 3, 365-385.

McKean, H.P., 1969. Stochastic Integrals, Academic Press, New York, 140 pp.

Müller, Claus, 1966. Spherical Harmonics - Lecture Notes in Mathematics, 17, Springer-Verlag, New York, 45 pp.

Neudecker, H., 1968. "The Kronecker matrix product and some of its applications in econometrics," Statistica Neerlandica, 22, 69-82.

Okamoto, M., 1973. "Distinctness of the Eigenvalues of a quadratic form in a multivariate sample," Ann. Statist., 1, 763-765.

Owens, W.H. 1973. "Strain modification of angular density distributions," Techtonophysics, 16, 249-261.

Pearson, K., 1905. "The problem of the random walk," Nature, 72, 294.

Pearson, Karl, 1906. "A mathematical theory of random migration" in Mathematical Contributions to the Theory of Evolution, XV" Draper's Company Research Memoirs, Biometric Series III, London.

Pitman, J., Yor, M., 1981. "Bessel processes and infinitely divisible laws," unpublished report, University of California, Berkeley.

Rao, C.R., 1973. Linear Statistical Inference and its Applications, John Wiley & Sons, New York, 625 pp.

Rayleigh, Lord, 1880. "On the resultant of a large number of vibrations of the same pitch and of arbitrary phase," Phil. Mag., (5), 10, 73-78.

Rayleigh, Lord, 1905. "The problem of random walk," Nature, 72, 318.

Rayleigh, Lord, 1919. "On the problem of random vibrations, and of random flights in one, two or three dimensions," Phil. Mag., (6), 37, 321-347.

Roberts, P.H., Ursell, H.D., 1960. "Random walk on the sphere and on a Riemannian manifold," Phil. Trans. Roy. Soc., A252, 317-356.

Ross, Ronald, 1923. Memoirs, E.P. Dutton & Co., New York, 547 pp.

Sander, B., 1930. "Gefugekunde der Gesteine," J. Springer, Vienna.

Saw, J.G., 1978. "A family of distributions on the m-sphere and some hypothesis tests," Biometrika, 65, 69-73.

Saw, J.G., 1981. "On solving the likelihood equations which derive from the Von Mises distribution," Technical Report, University of Florida.

Scheidegger, A.E., 1965. "On the statistics of the orientation of bedding planes, grain axes and similar sedimentological data," U.S. Geol. Survey Prof. Paper, 525-C, 164-167.

Schmidt-Koenig, K., 1972. "New experiments on the effect of clock shifts on homing pigeons in animal orientation and navigation," Eds: S.R. Galler, K. Schmidt-Koenig, G.J. Jacobs and R.E. Belleville, NASA SP-262, Washington, D.C.

Selby, B., 1964. "Girdle distributions on the sphere," Biometrika, 51, 381-392.

Sibuya, M., 1962. "A method for generating uniformly distributed points on n-dimensional spheres," Ann. Inst. Statist. Math., 14, No. 1, 81-85.

Stam, A.J., 1982. "Limit theorems for uniform distributions on spheres in high dimensional euclidean spaces," J. Appl. Prob., 19, No. 1, 221-229.

Stephens, M.A., 1963. "Random walk on the circle," Biometrika, 50, 385-390.

Stephens, M.A., 1964. "The testing of unit vectors for randomness," J. Amer. Statist. Soc., 59, 160-167.

Stephens, M.A., 1969. "Tests for randomness of directions against two circular alternatives," J. Amer. Statist. Ass., 64, 280-289 .

Szegö, G., 1939. Orthogonal Polynomials, 23, American Mathematical Society Collegquium Publications, Providence, 432 pp.

Tapia, R.A., Thompson, J.R., 1978. Nonparametric Probability Density Estimation, The Johns Hopkins University Press, Baltimore and London, 176 pp.

Tashiro, Y., 1977. "On methods for generating uniform random points on the surface of a sphere," Ann. Inst. Statist. Math, 29, Part A, 295-300.

Teicher, H., 1961. "Maximum likelihood characterization of distribution," Ann. Math. Statist., 32, 1214-1222.

Van Alstine, D.R., 1979. "Apparent polar wandering with respect to North America since the late Precambrian," Ph.D. Thesis, Calif. Inst. of Tech., Pasadena.

Von Mises, R., 1981. "Uber die "Ganzzahligkeit" der Atomgewicht und Verwandte Fragen," Physikal. Z., 19, 490-500.

Vincenz, S.A., Bruckshaw, J. McG., 1960. "Note on the probabilities distribution of a small number of vectors," Proc. Camb. Phil. Soc., 56, 21-26.

Watson, G.S., Williams, E.J., 1956. "On the construction of significance tests on the circle and the sphere," Biometrika, 43, Parts 3 and 4, 345-352.

Watson, G.S., 1956. "Analysis of dispersion on a sphere," Monthly Notices Roy. Astro. Soc. Geophys. Suppl., 7, 4, 153-159.

Watson, G.S., 1956. "A test for randomness of directions," Monthly Notices Roy. Astro. Soc. Geophys. Suppl., 7, 4, 160-161.

Watson, G.S., Irving, E., 1957. "Statistical methods in rock magnetism," Monthly Notices Roy. Astro. Soc., 7, 6, 290-300.

Watson, G.S., 1960. "More significance tests on the sphere," Biometrika, 47, Parts 1 and 2, 87-91.

Watson, G.S., 1961. "Goodness-of-fit tests on a circle," Biometrika, 48, Parts 1 and 2, 109-114.

Watson, G.S., 1962. "Goodness-of-fit tests on a circle-II," Biometrika, 49, Parts 1 and 2, 57-63.

Watson, G.S., Leadbetter, M.R., 1963. "On the estimation of the probability density-I," Ann. Math. Stat., 34, 2, 480-491.

Watson, G.S., Wheeler, S., 1964. "A distribution-free two-sample test on a circle," Biometrika, 51, Parts 1 and 2, 256.

Watson, G.S., 1965. "Equatorial distributions on a sphere," Biometrika, 52, Parts 1 and 2, 193-201.

Watson, G.S., 1966. "Statistics of orientation data," Jour. of Geology, 74, 5, Part 2, 786-797. (Reprinted in "Statistical Analysis of Geology," Benchmark Papers in Geology 37, Dowden, Hutchinson, and Ross.)

Watson, G.S., 1967. "Another test for the uniformity of a circular distribution," Biometrika, 54, Parts 3 and 4, 675-677.

Watson, G.S., Beran, R.J., 1967. "Testing a sequence of unit vectors for serial correlation," Jour. of Geophysical Research, 72, 22, 5655-5659.

Watson, G.S., 1967. "Some problems in the statistics of directions," Bull. of I.S.I., (36th Session of I.S.I, Australia) 42, 374-385.

Watson, G.S., 1968. "Orientation statistics in the earth sciences," Bull. of the Geological Institutions of the Univ. of Uppsala, New Series, 2, 9, 73-89. (Symposium on Biometry in Paleontology.)

Watson, G.S., 1969. "Density estimation by orthogonal series," Ann. Math Stat., 40, 4, 1469-1498.

Watson, G.S., 1970. "The statistical treatment of orientation data," Geostatistics - a colloquium (Ed., D.F. Merriam), Plenum Press, New York, 1-10.

Watson, G.S., Epp, R., Tukey, J.W., 1971. "Testing unit vectors for correlation," J. of Geophysical Research, 76, 35, 8480-8483.

Watson, G.S., 1974. "Optimal invariant tests for uniformity," in "Studies in Probability and Statistics," Editor: E.J. Williams, Jerusalem Academic Press, 121-128.

Watson, G.S., 1981. "The theory of concentrated Langevin distributions," to appear in J. Mult. Anal.

Watson, G.S., 1981. "Large sample theory of the Langevin distributions," to appear in J. Stat. Planning Inference.

Watson, G.S., 1982. "Asymptotic spectral analysis of cross-product matrices," to appear in "Specification Analysis of the Linear Model," a book.

Watson, G.S., 1982. "Large sample theory for distributions on the hypersphere with rotational symmetries, to appear in J. Inst. Math. Stat.

Watson, G.S., 1982. "The estimation of palaeomagnetic pole positions, pp. 703-712 in Statistics in Probability: Essay in honor of C.R. Rao, North-Holland, Amsterdam and New York.

Watson, G.S., 1982. "Distributions on the circle and sphere," pp. 265-280 in Essays in Statistical Science, J. App. Prob. Special Volume 19A, published by Applied Probability Trust.

Watson, G.S., 1982. "Limit Theorems on high dimensional spheres and Stiefel Manifolds," to appear in volume in honor of T.W. Anderson.

Watson, G.S., 1982. "Distributions on the circle and sphere," pp. 265-280 in Essays in Statistical Science, J. App. Prob. Special Volume 19A, published by Applied Probability Trust.

Wellner, J., 1978. "Two sample tests for a class of distributions on the sphere," unpublished report, University of Rochester.

Wellner, J., 1979. "Permutation tests for directional data," Ann. Statist., 7, 924-943 pp.